TO THE STUDENT

The Student Study Art Notebook is designed to help you in your study of microbiology. The notebook contains art taken directly from the text and overhead transparencies; thus you can take notes during the lectures, or jot down comments as you are reading through the chapters.

The notebook is 3-hole punched so you can remove sheets and put them in a binder with other study or lecture notes. Any blank pages at the end of this notebook can be used for additional notes or drawings.

We hope this notebook, used along with your text, helps to make the study of microbiology easier for you.

student study
ART NOTEBOOK

Microbiology
Third Edition
2003

Lansing M. Prescott
Augustana College

John P. Harley
Eastern Kentucky University

Donald A. Klein
Colorado State University

WCB
Wm. C. Brown Publishers

Dubuque, IA   Bogotá   Buenos Aires   Caracas   Chicago   Guilford, CT   London
Madrid   Mexico City   Seoul   Singapore   Sydney   Taipei   Tokyo   Toronto

A Times Mirror Company

The credits section for this book begins on page 113 and
is considered an extension of the copyright page.

ISBN 0–697–21874–0

Printed in the United States of America by Wm. C. Brown Communications, Inc.,
2460 Kerper Boulevard, Dubuque, IA 52001

10 9 8 7 6 5 4 3 2 1

DIRECTORY OF NOTEBOOK FIGURES

TO ACCOMPANY PRESCOTT ET AL.
MICROBIOLOGY, 3E.

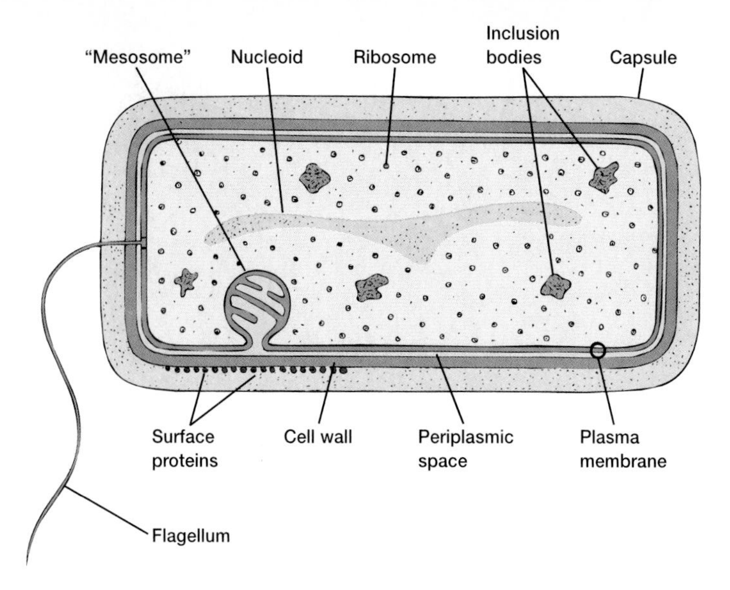

Morphology of Gram-Positive Cell
Figure 3.4

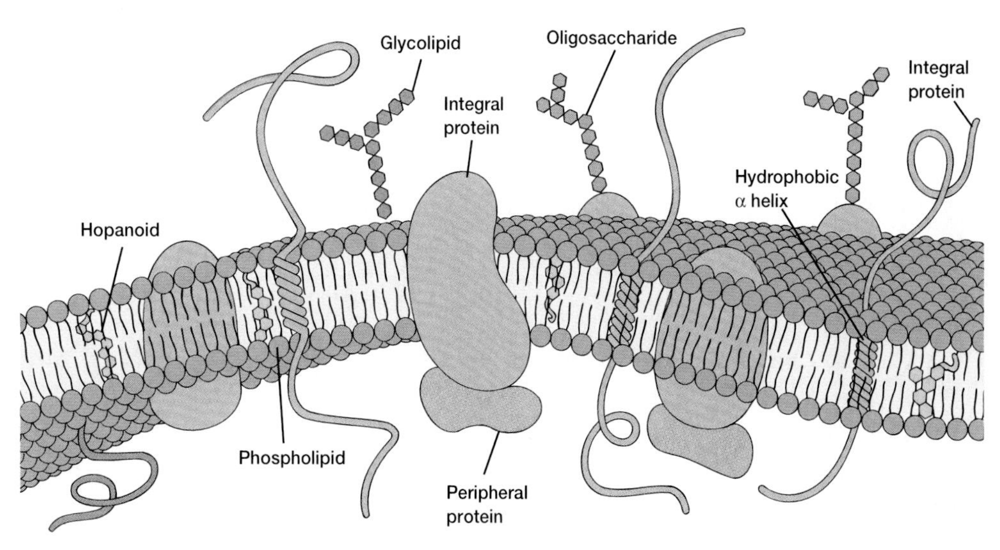

Plasma Membrane Structure
Figure 3.7

1

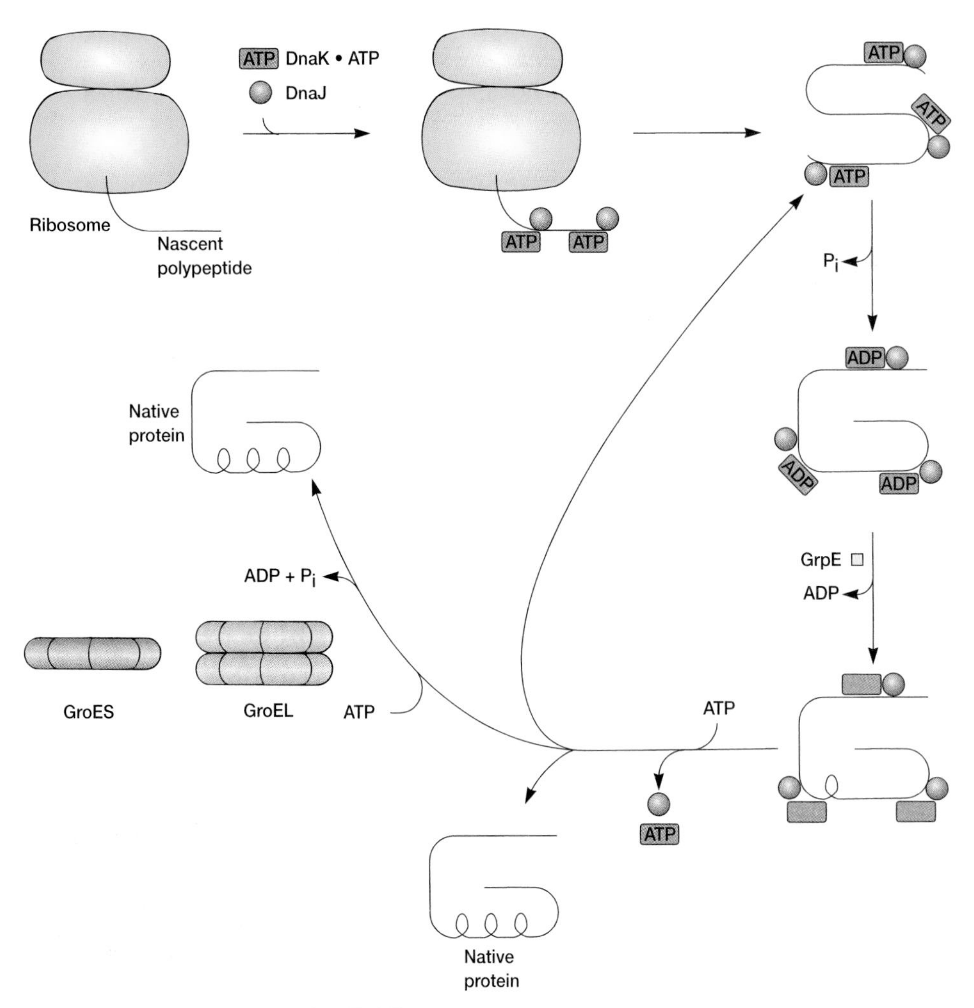
ATP DnaK • ATP
DnaJ
Ribosome
Nascent polypeptide
ATP
ATP
ATP
ATP
ATP
P_i
ADP
ADP
ADP
GrpE
ADP
Native protein
ADP + P_i
GroES
GroEL
ATP
ATP
ATP
Native protein

Chaperones and Polypeptide Folding
Figure 3.16

NAM   NAG

CH₂OH   H   NH─C═O
                          |
      H   O   CH₃
                   OH
─O─   H   O─   H   H
      H   H   ─O─
                         O
      H   NH─C═O   CH₂OH
              |
              CH₃   D–Lactic acid
      O
H₃C─CH─C═O }
      NH
              } L–Alanine
CH₃─C─H
      |
      C═O
      |
      NH
      |
H─C─CH₂─CH₂─C═O } D-Glutamic acid
      |
      COOH

      NH
      |
H─C─(CH₂)₃─CH─COOH } *meso*-Diaminopimelic acid
      | |
      C═O   NH₂
      |
      NH
      |
H─C─CH₃ } D-Alanine
      |
      C═O
      |
      ↖ May be connected to the peptide interbridge
        or to the diaminopimelic acid in another tetrapeptide chain

Peptidoglycan Subunit Composition
Figure 3.19

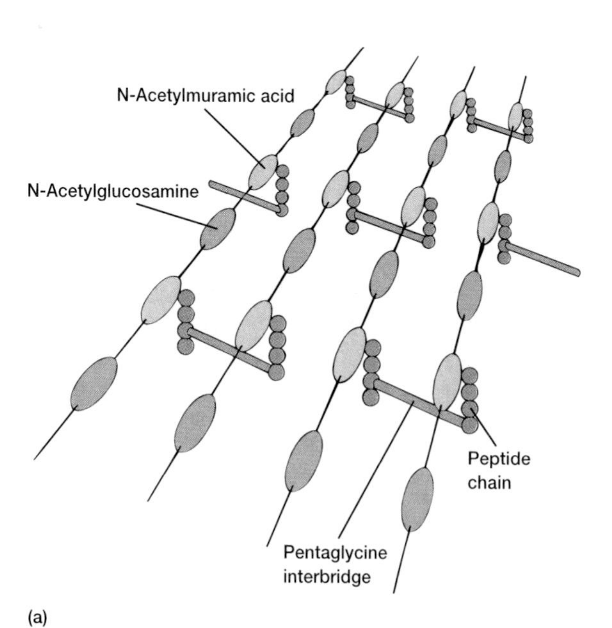

N-Acetylmuramic acid

N-Acetylglucosamine

Peptide
chain

Pentaglycine
interbridge

(a)

Peptidoglycan Structure
Figure 3.22 *a*

3

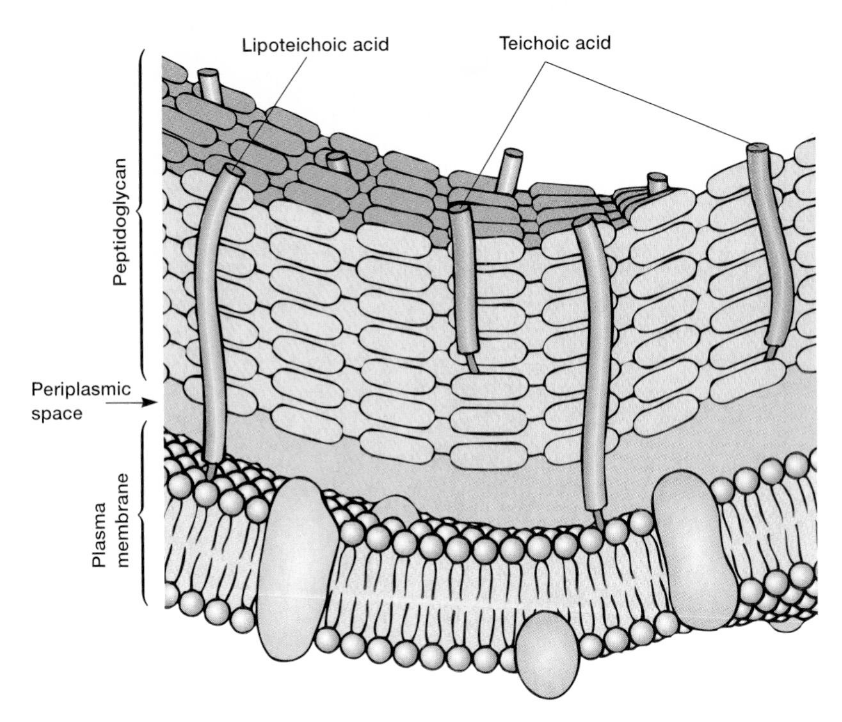

Gram-Positive Envelope
Figure 3.24

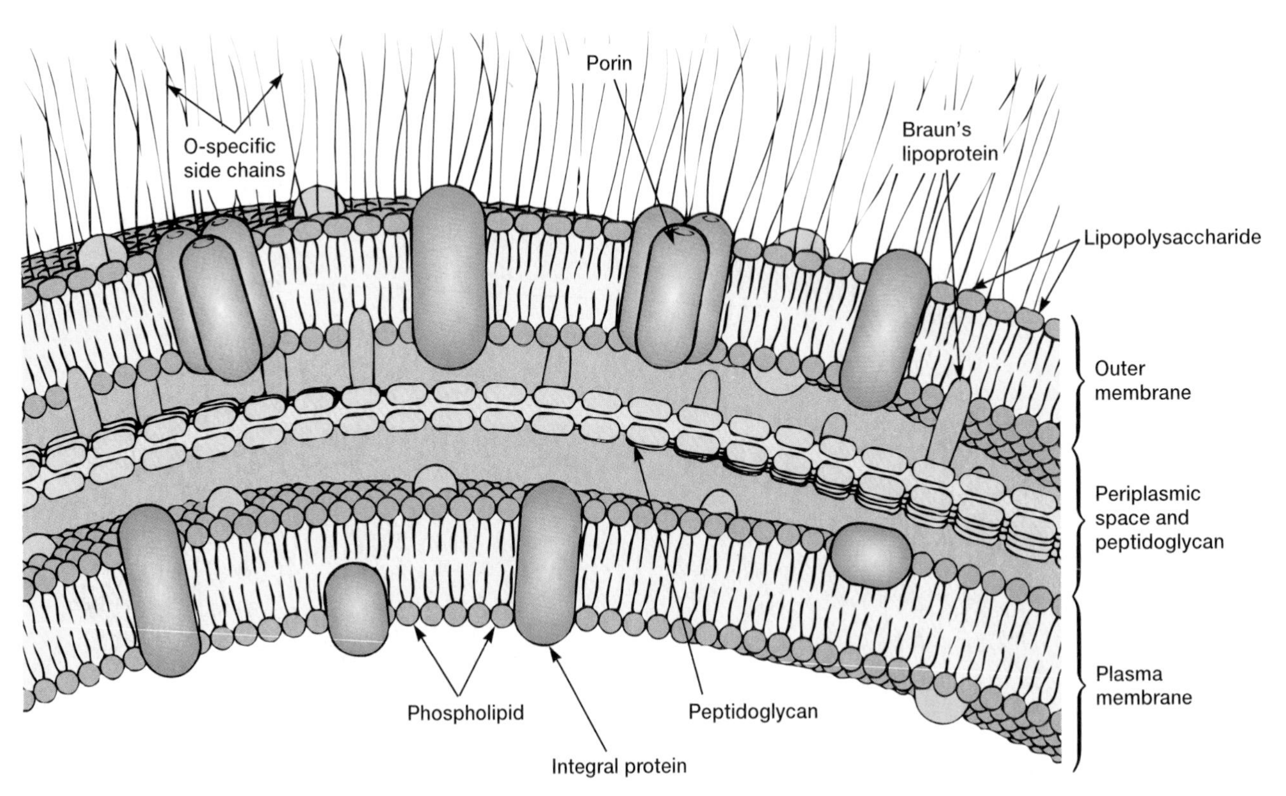

Gram-Negative Envelope
Figure 3.26

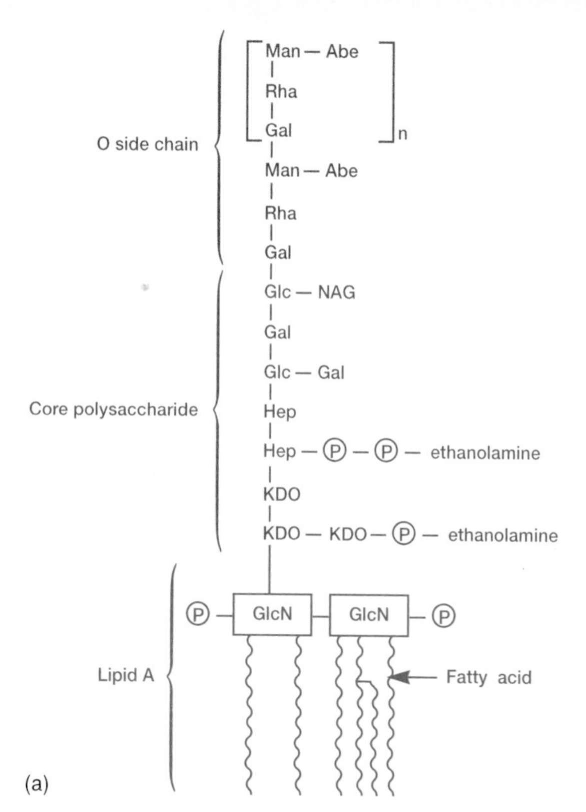

Lipopolysaccharide Structure
Figure 3.27 *a*

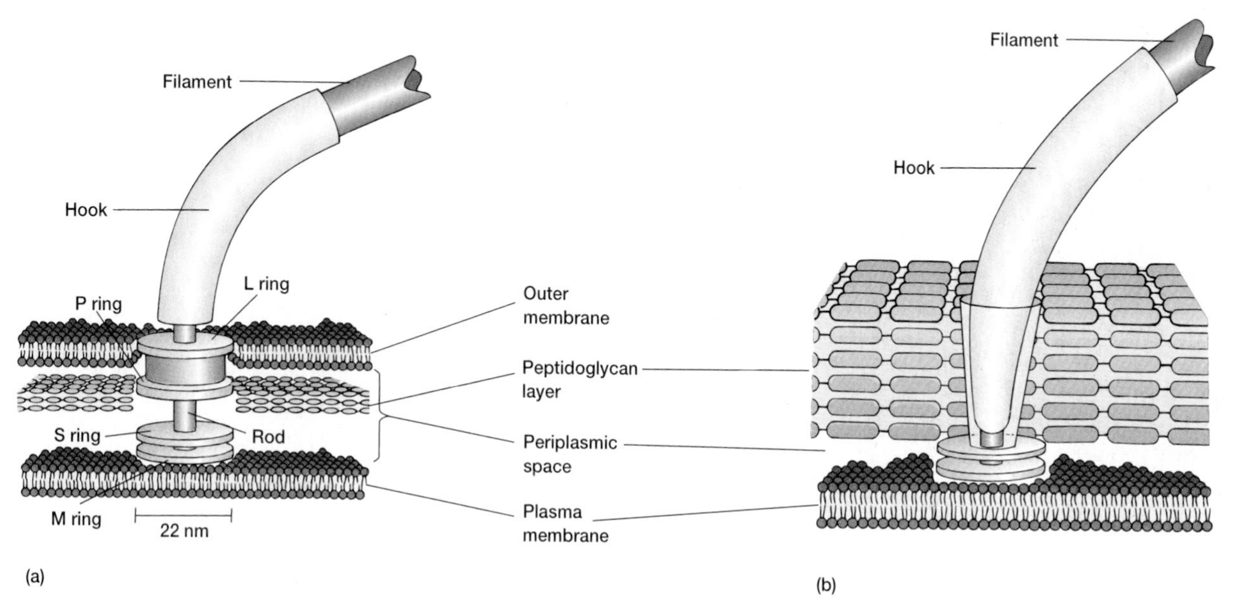

Ultrastructure of Bacterial Flagella
Figure 3.35

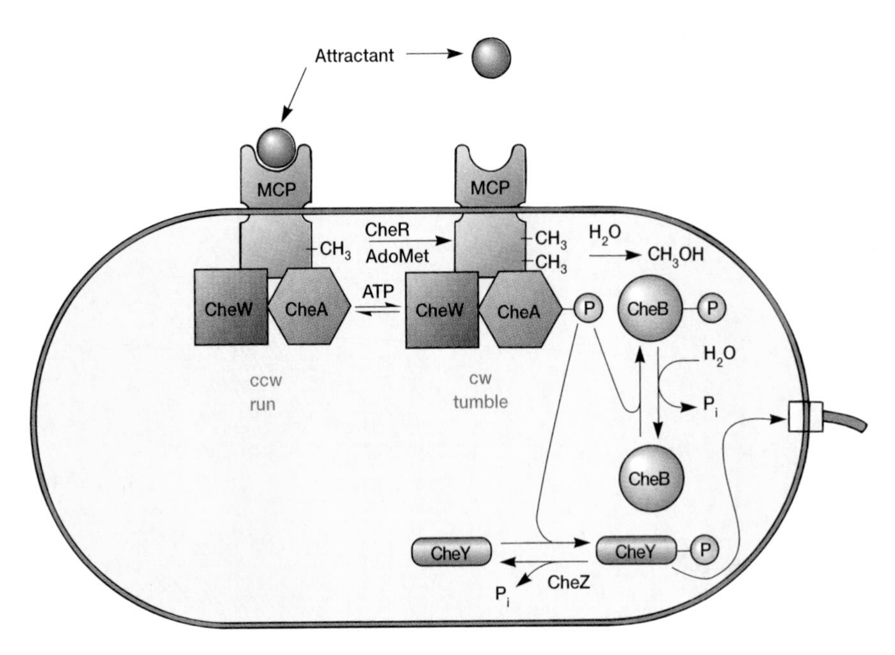
Attractant
MCP
MCP
$-CH_3$
CheR
AdoMet
$-CH_3$
$-CH_3$
H_2O
CH_3OH
CheW
CheA
ATP
CheW
CheA
P
CheB
P
H_2O
P_i
ccw
run
cw
tumble
CheB
CheY
CheY
P
P_i
CheZ

Mechanism of Chemotaxis in *E. coli*
Figure 3.41

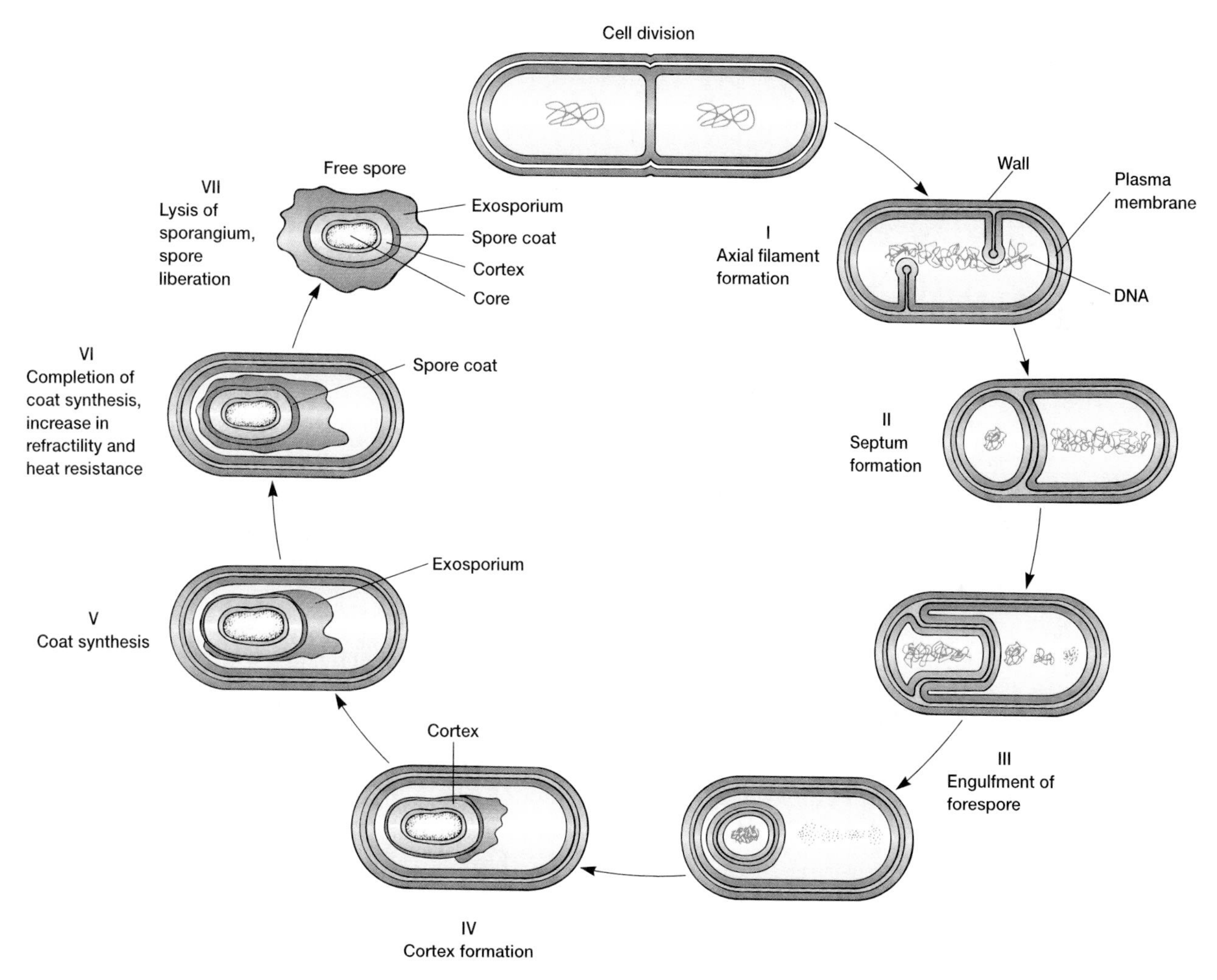

Cell division

I Axial filament formation

Wall

Plasma membrane

DNA

II Septum formation

III Engulfment of forespore

IV Cortex formation

Cortex

V Coat synthesis

Exosporium

VI Completion of coat synthesis, increase in refractility and heat resistance

Spore coat

VII Lysis of sporangium, spore liberation

Free spore

Exosporium
Spore coat
Cortex
Core

Endospore Formation
Figure 3.45

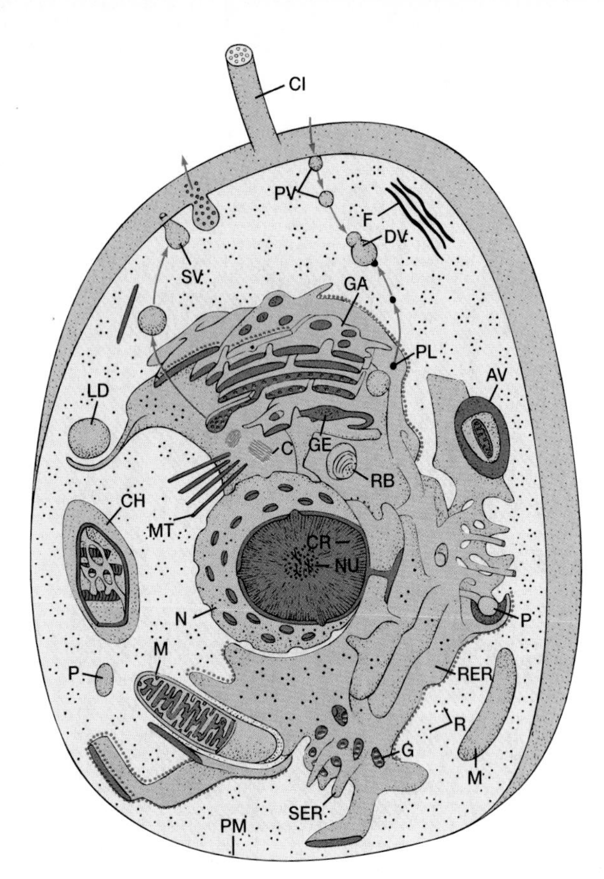

Eucaryotic Cell Ultrastructure
Figure 4.3

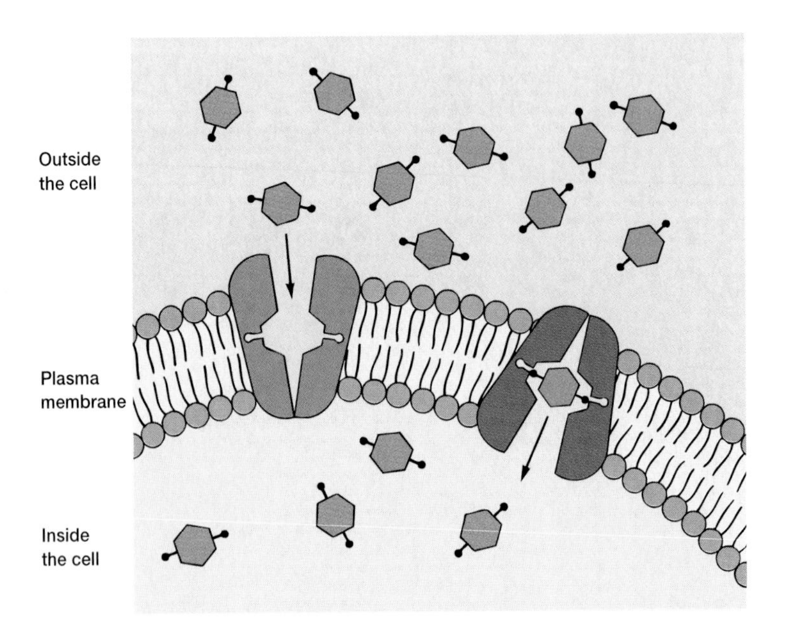

Outside
the cell

Plasma
membrane

Inside
the cell

Model of Facilitated Diffusion
Figure 5.2

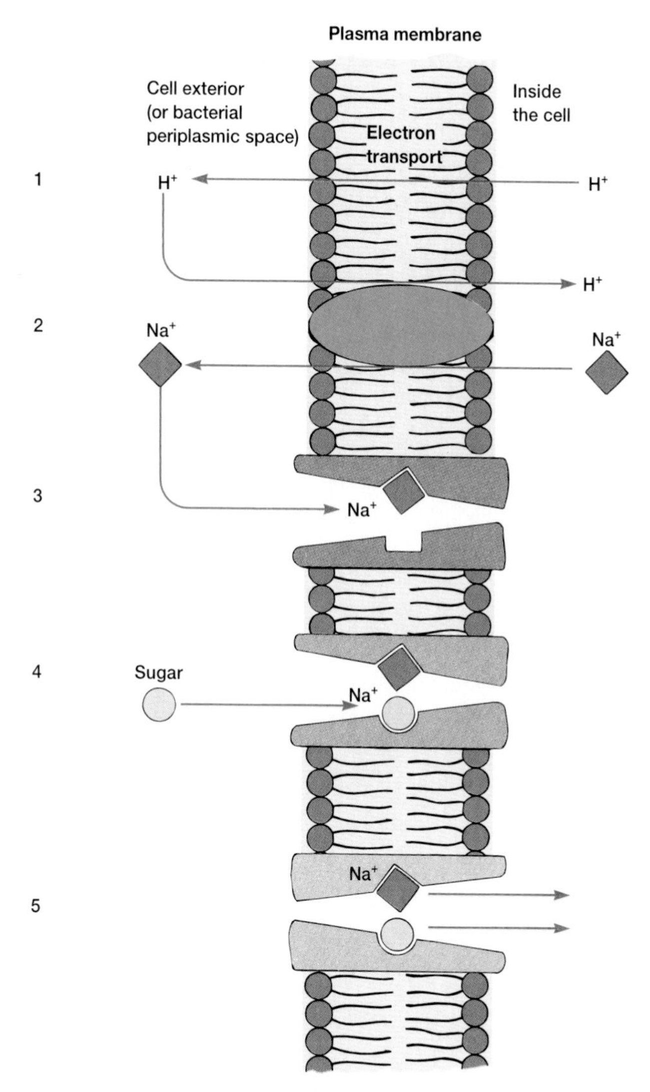
Plasma membrane
Cell exterior
(or bacterial
periplasmic space)
Inside
the cell
Electron
transport
1
H⁺
H⁺
H⁺
2
Na⁺
Na⁺
3
Na⁺
4
Sugar
Na⁺
5
Na⁺

Active Transport
Figure 5.3

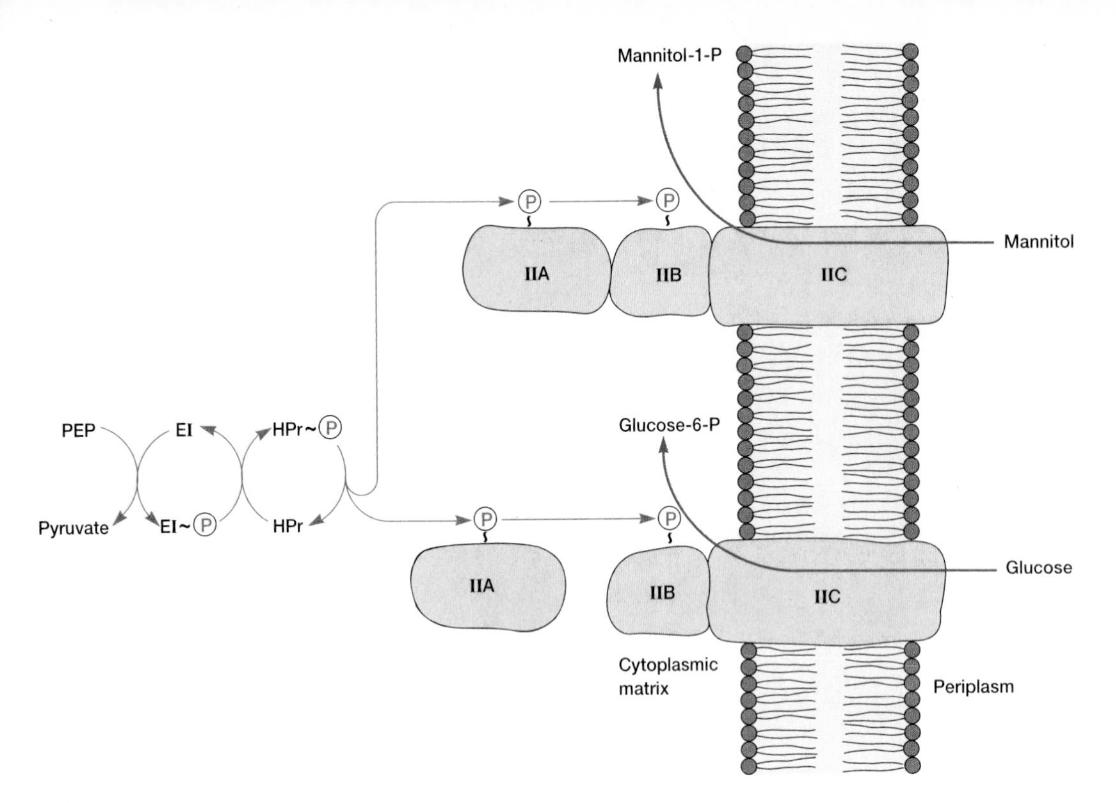

Bacterial PTS Transport
Figure 5.4

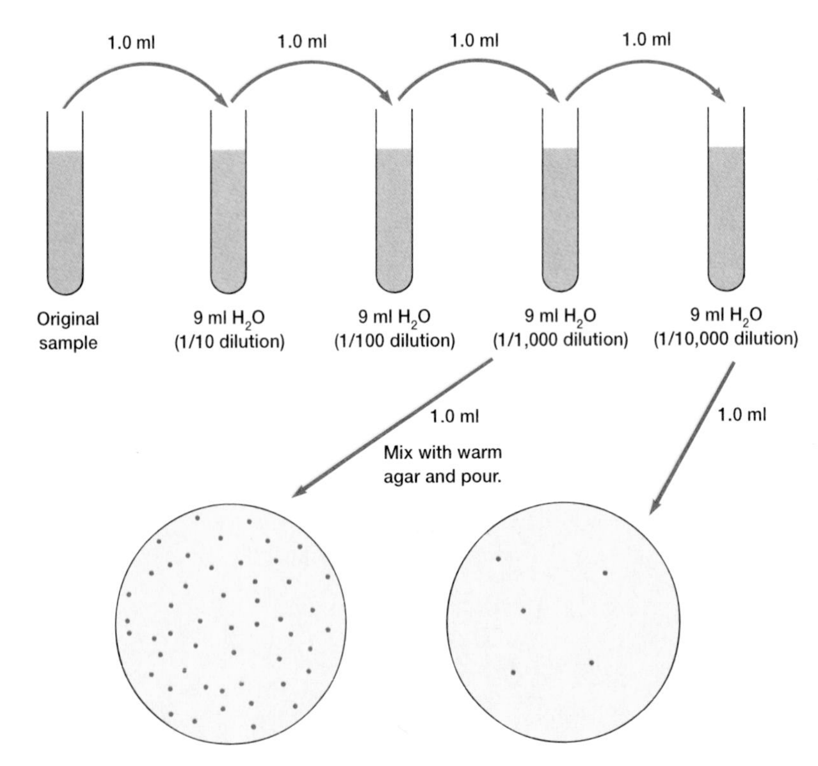

The Pour-Plate Technique
Figure 5.9

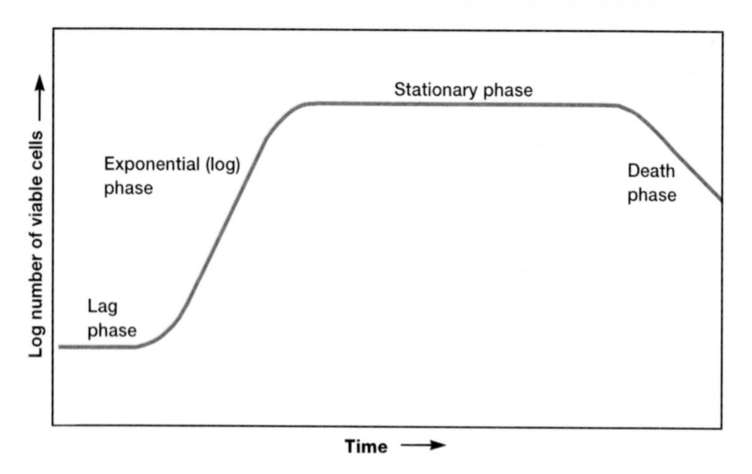

Microbial Growth Curve
Figure 6.1

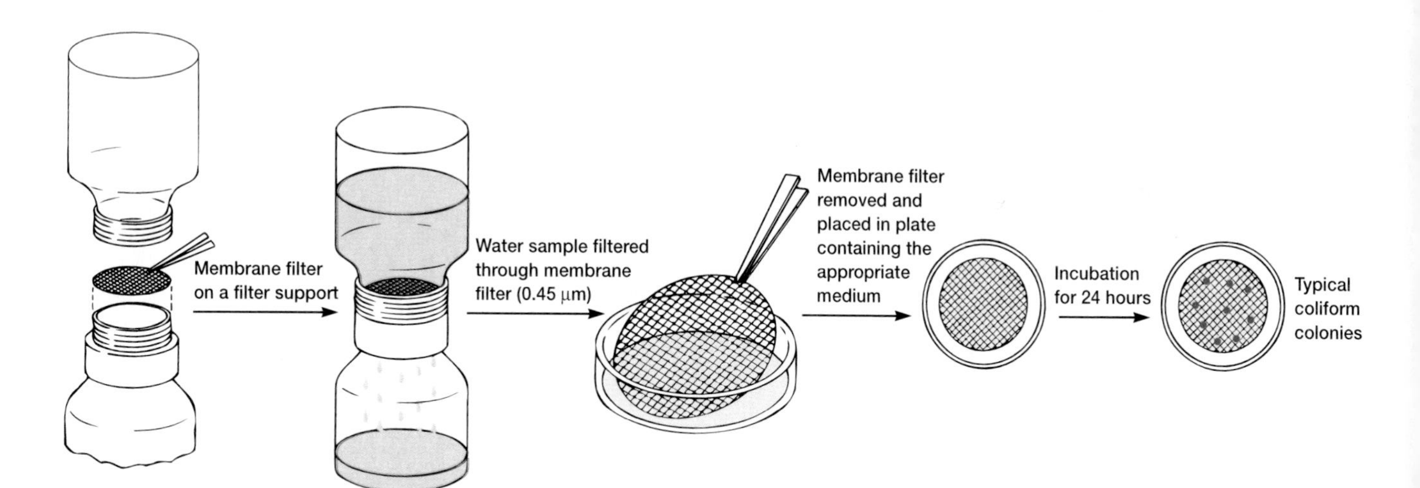

The Membrane Filtration Procedure
Figure 6.7

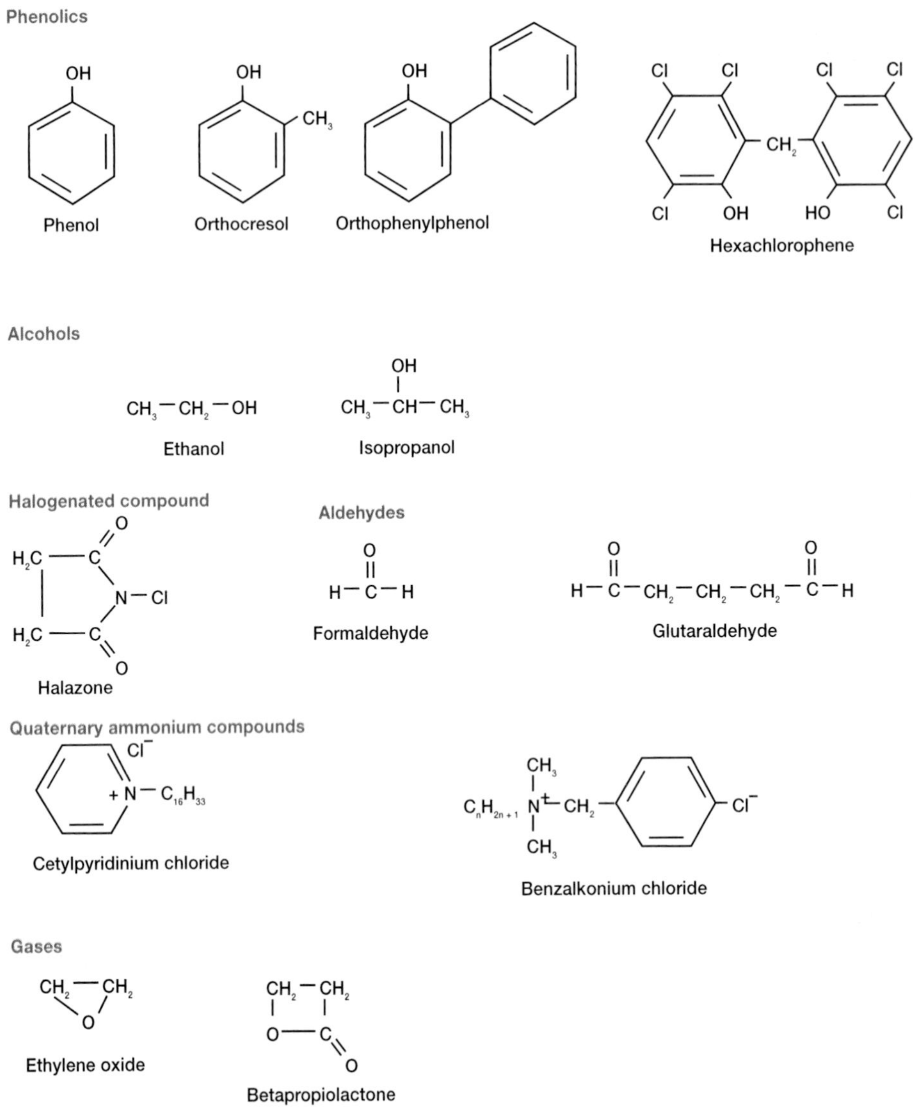

Phenolics

Phenol · Orthocresol · Orthophenylphenol · Hexachlorophene

Alcohols

$CH_3 - CH_2 - OH$
Ethanol

$CH_3 - \overset{\displaystyle OH}{\underset{|}{CH}} - CH_3$
Isopropanol

Halogenated compound

Halazone

Aldehydes

Formaldehyde

Glutaraldehyde

Quaternary ammonium compounds

Cetylpyridinium chloride

Benzalkonium chloride

Gases

Ethylene oxide

Betapropiolactone

Disinfectants and Antiseptics
Figure 7.7

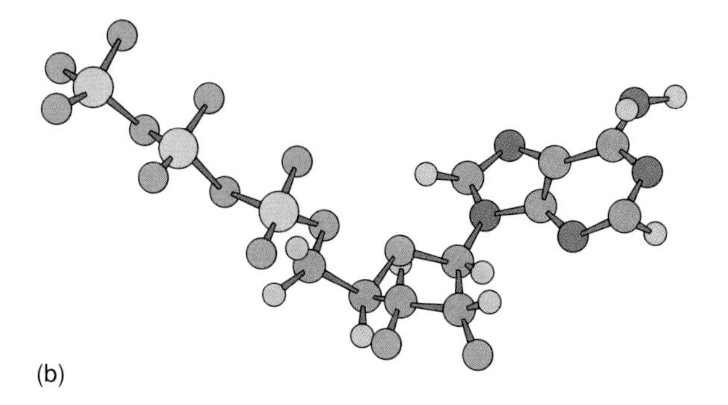

(a)

Adenosine

Adenosine diphosphate (ADP)

Adenosine triphosphate (ATP)

(b)

ATP and ADP
Figure 8.2

(a)

Nicotinamide

Ribose

Adenine

Ribose

NADP has a phosphate here.

(b)

Reduced substrate

NAD⁺

Oxidized substrate

NADH

Structure and Function of NAD
Figure 8.8 *a, b*

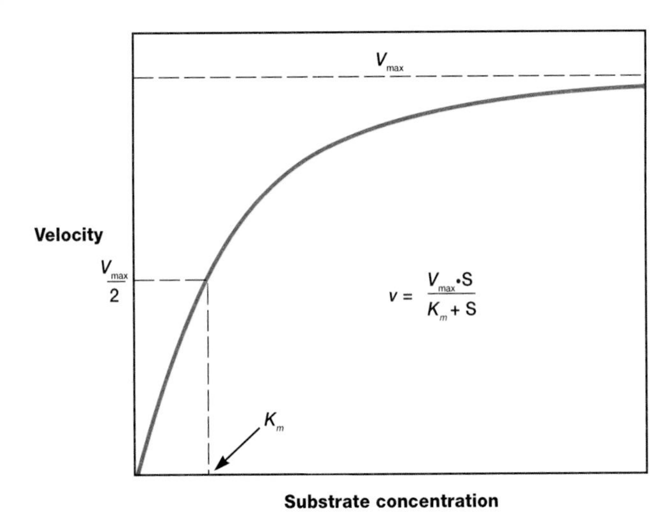

Michaelis-Menten Kinetics
Figure 8.16

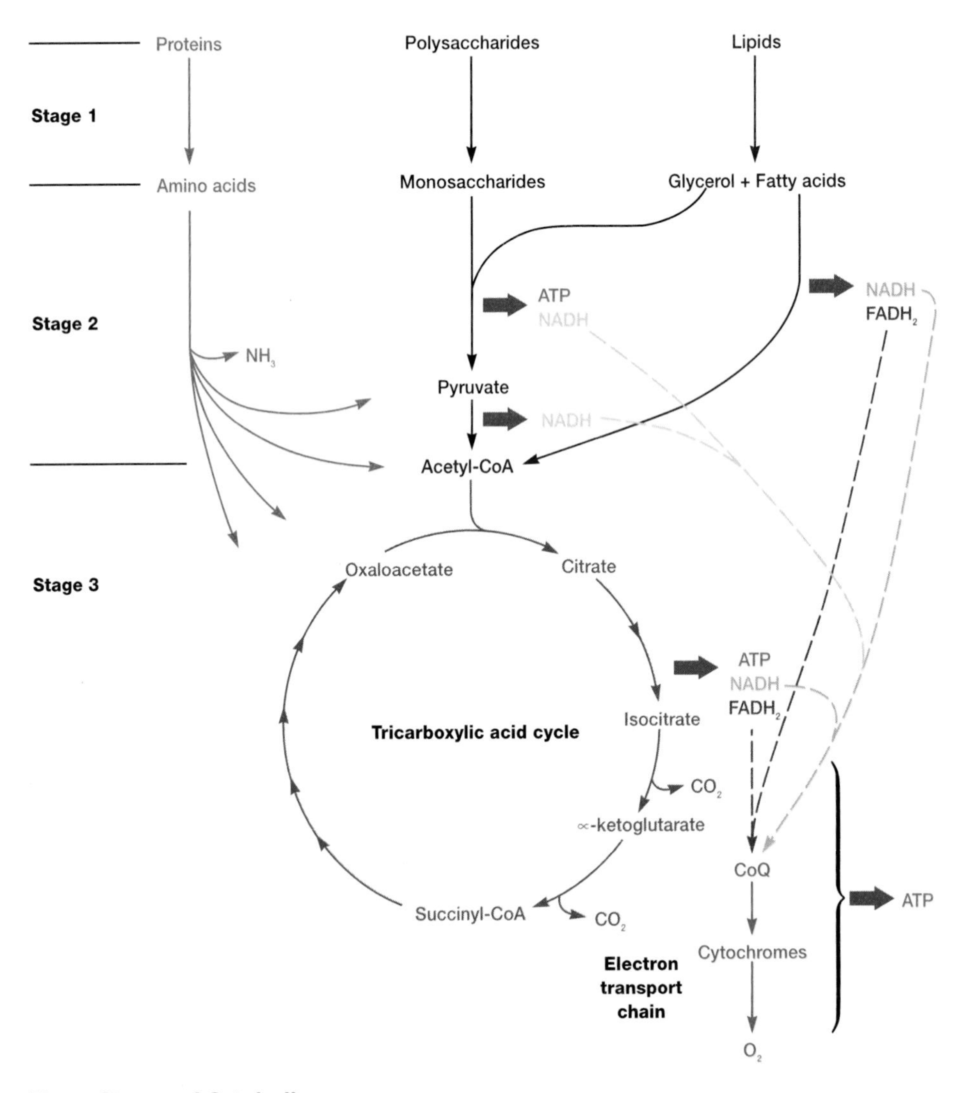

Three Stages of Catabolism
Figure 9.1

14

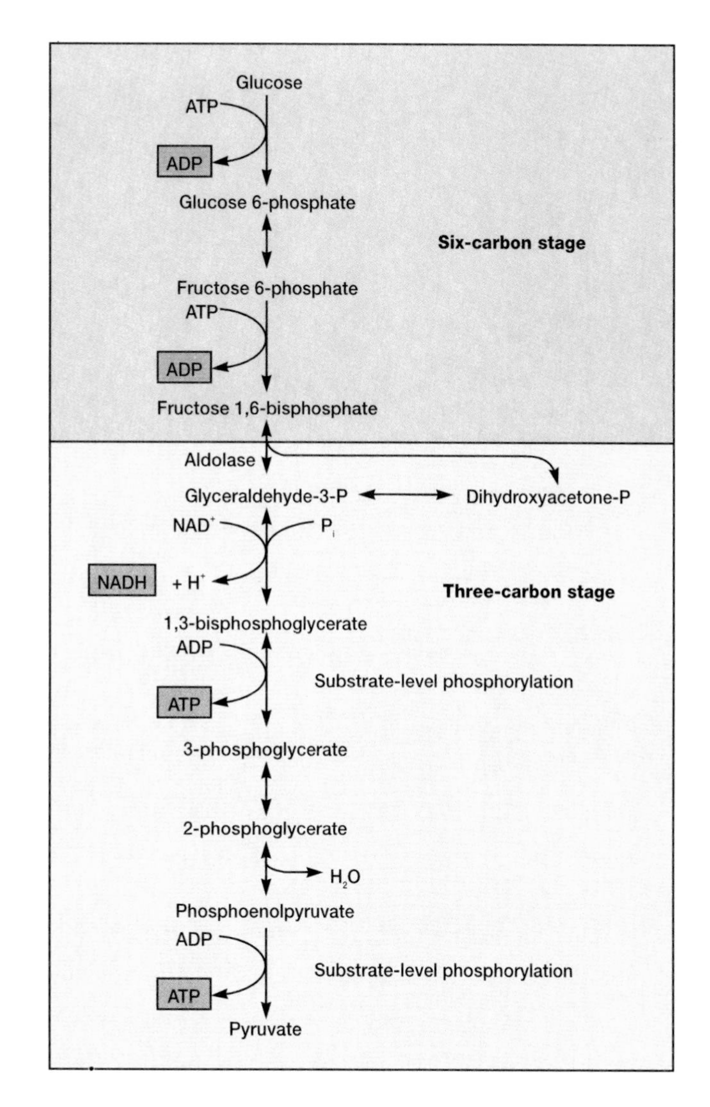
Glucose
ATP
ADP
Glucose 6-phosphate
Six-carbon stage
Fructose 6-phosphate
ATP
ADP
Fructose 1,6-bisphosphate
Aldolase
Glyceraldehyde-3-P
Dihydroxyacetone-P
NAD^+
P_i
NADH
$+ H^+$
Three-carbon stage
1,3-bisphosphoglycerate
ADP
Substrate-level phosphorylation
ATP
3-phosphoglycerate
2-phosphoglycerate
H_2O
Phosphoenolpyruvate
ADP
Substrate-level phosphorylation
ATP
Pyruvate

Glycolysis
Figure 9.3

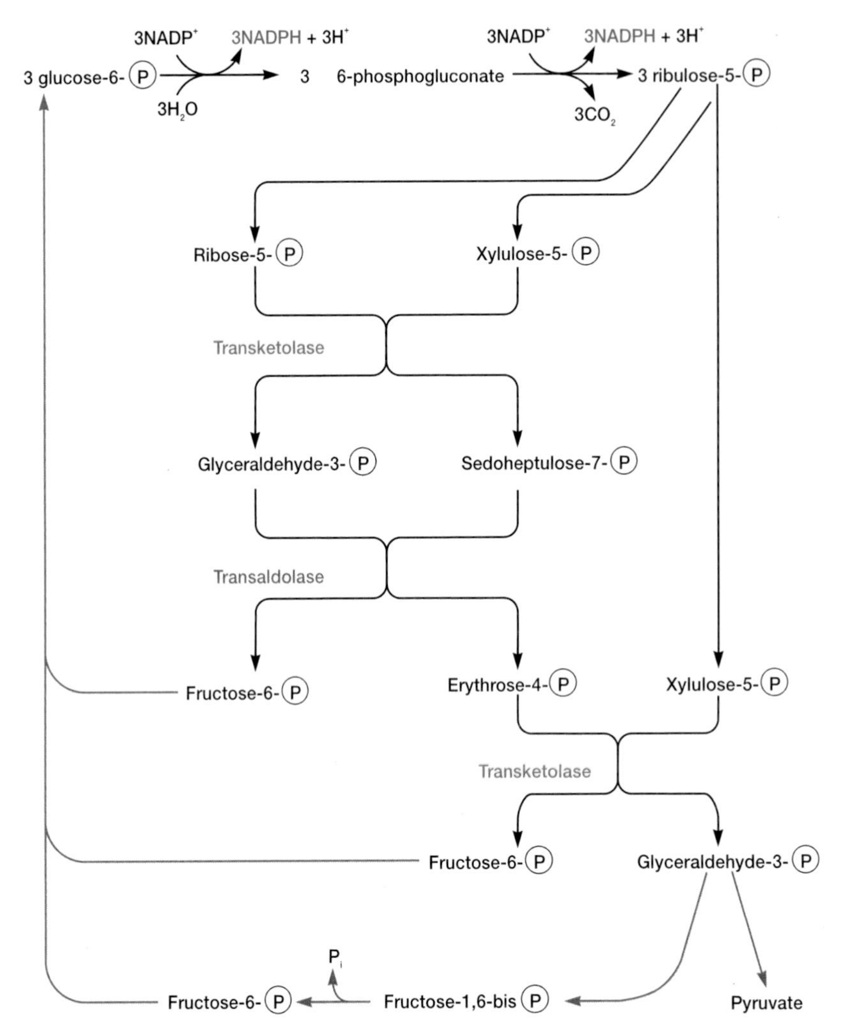

3NADP+
3NADPH + 3H+
3 glucose-6-(P)
3 6-phosphogluconate
3H2O
3NADP+
3NADPH + 3H+
3 ribulose-5-(P)
3CO2
Ribose-5-(P)
Xylulose-5-(P)
Transketolase
Glyceraldehyde-3-(P)
Sedoheptulose-7-(P)
Transaldolase
Fructose-6-(P)
Erythrose-4-(P)
Xylulose-5-(P)
Transketolase
Fructose-6-(P)
Glyceraldehyde-3-(P)
Pi
Fructose-6-(P)
Fructose-1,6-bis (P)
Pyruvate

Pentose Phosphate Pathway
Figure 9.4

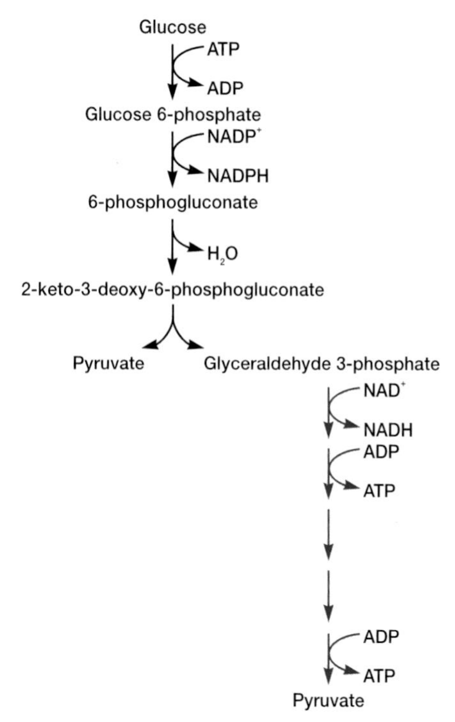

Entner-Doudoroff Pathway
Figure 9.6

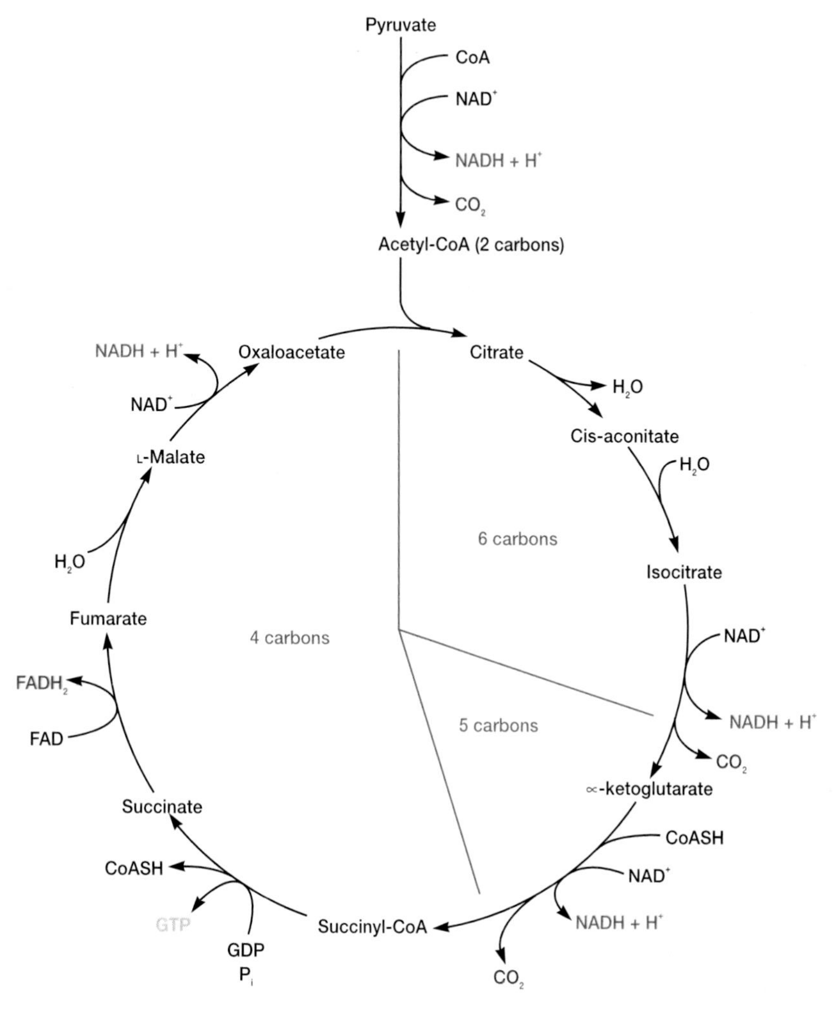

Tricarboxylic Acid Cycle
Figure 9.7

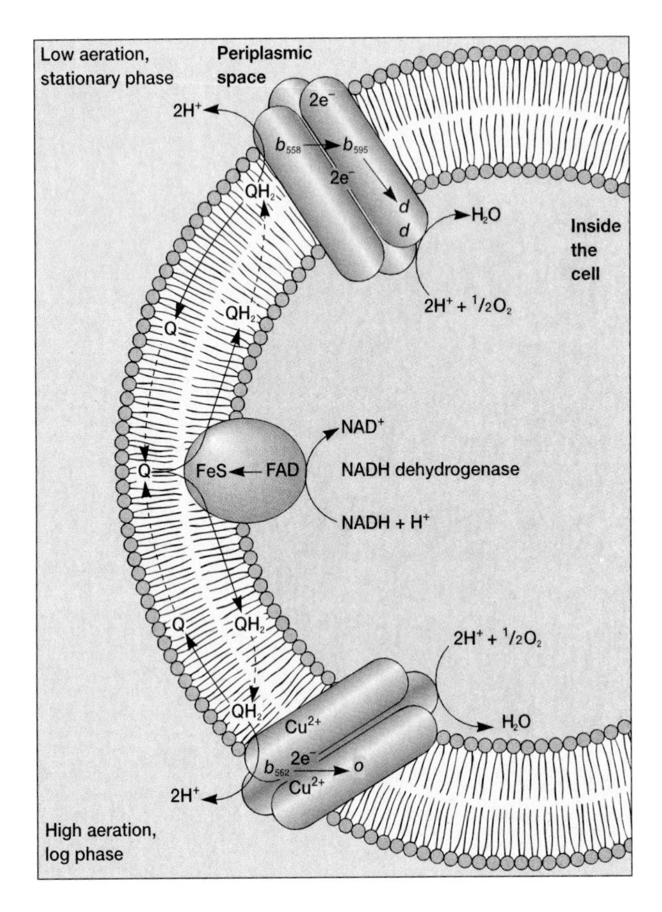

Low aeration,
stationary phase

Periplasmic
space

$2e^-$

$2H^+$

b_{558} → b_{595}

$2e^-$

QH_2

d
d

→ H_2O

Inside
the
cell

QH_2

Q

QH_2

$2H^+ + {}^1/{}_2O_2$

NAD$^+$

Q

FeS ← FAD

NADH dehydrogenase

NADH + H$^+$

Q

QH_2

$2H^+ + {}^1/{}_2O_2$

QH_2

Cu^{2+}

b_{562} $2e^-$ o

$2H^+$

Cu^{2+}

→ H_2O

High aeration,
log phase

The Aerobic Respiratory System of *E. coli*
Figure 9.9

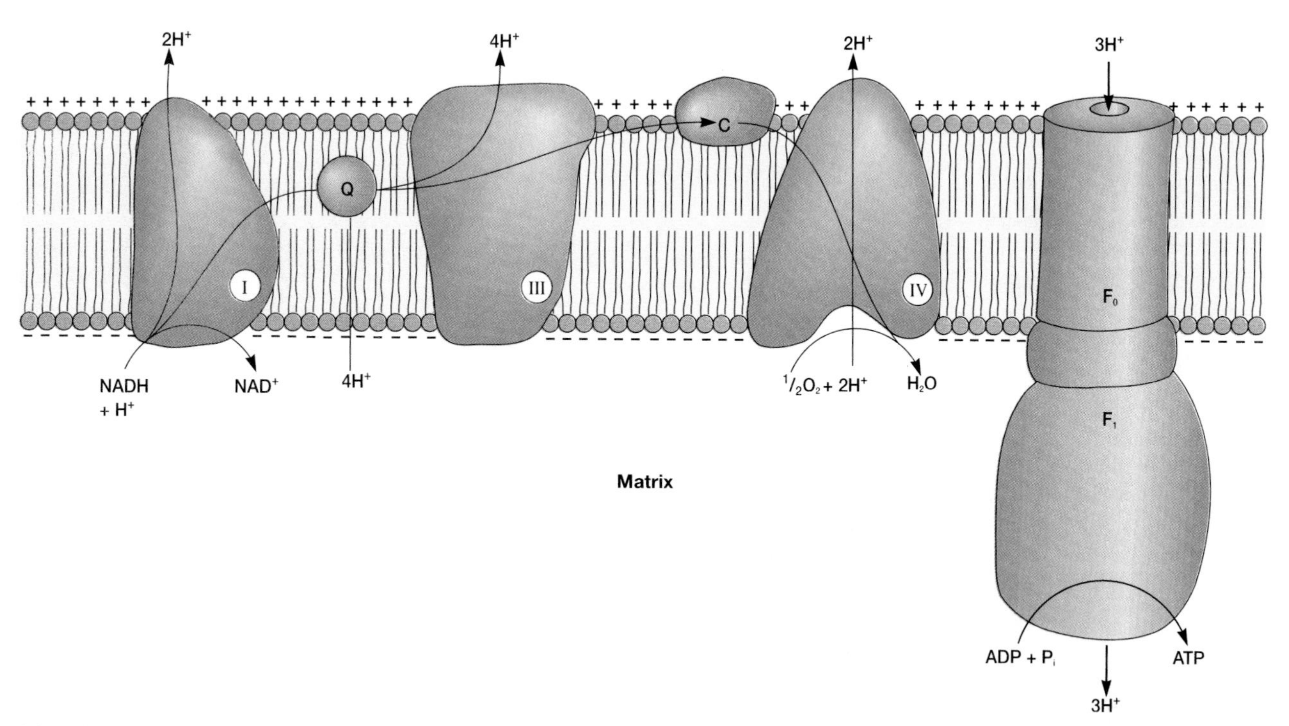

Chemiosmosis
Figure 9.10

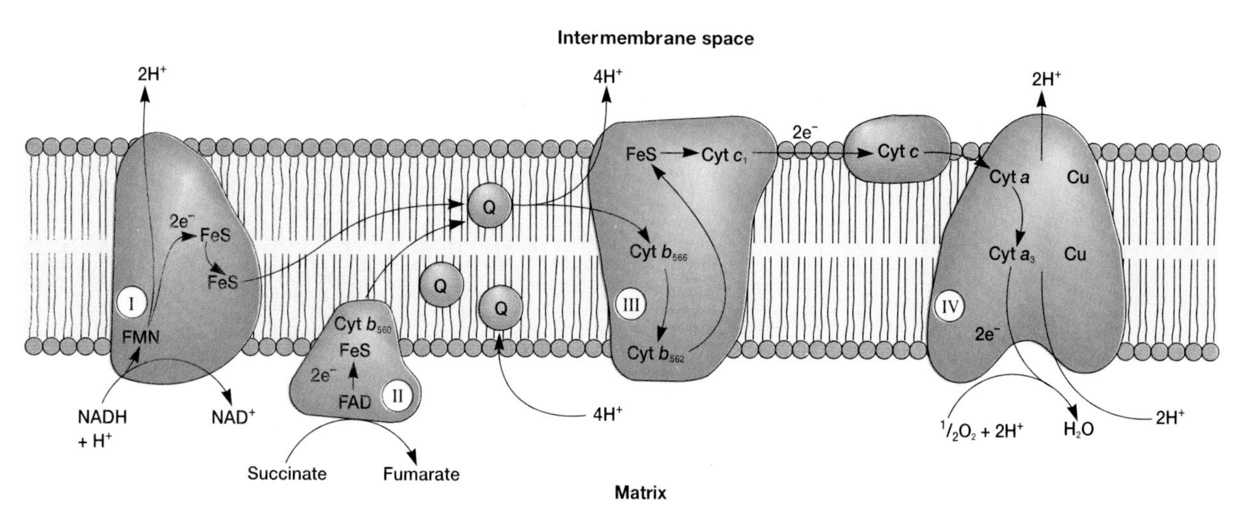

Chemiosmotic Hypothesis and Mitochondrion
Figure 9.11

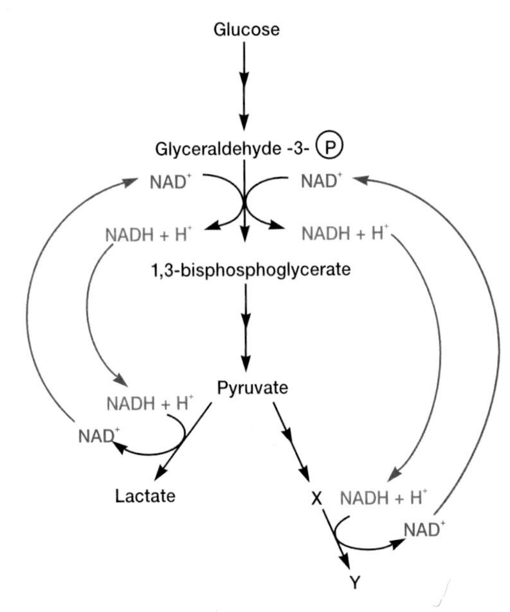

Reoxidation of NADH during Fermentation
Figure 9.13

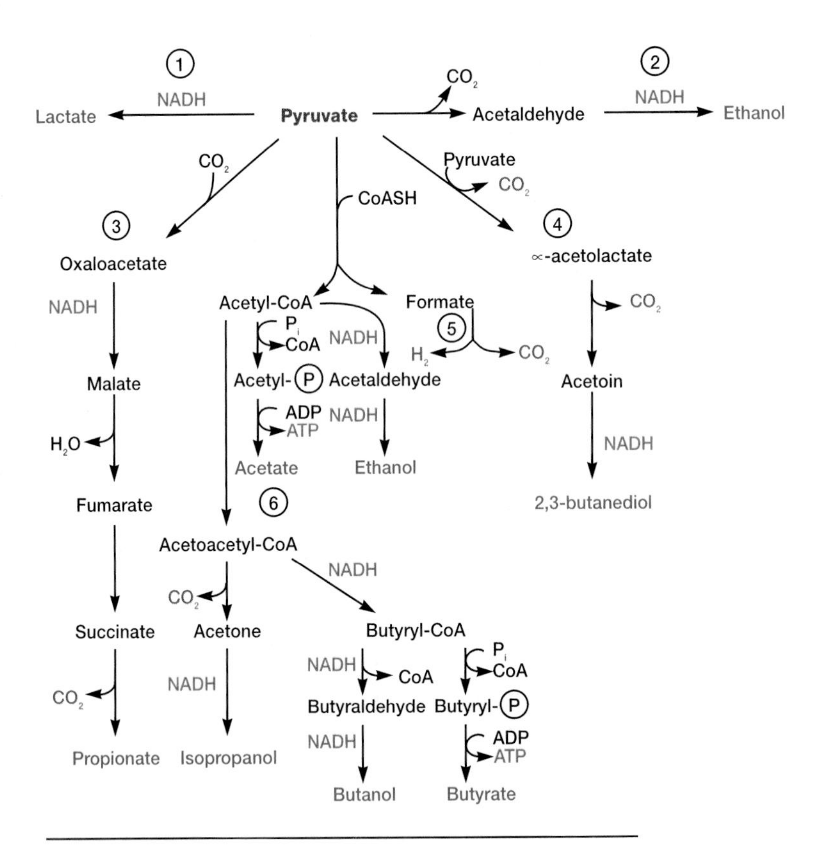

1. Lactic acid bacteria (*Streptococcus, Lactobacillus*), *Bacillus*

2. Yeast, *Zymomonas*

3. Propionic acid bacteria (*Propionibacterium*)

4. *Enterobacter, Serratia, Bacillus*

5. Enteric bacteria (*Escherichia, Enterobacter, Salmonella, Proteus*)

6. *Clostridium*

Some Common Microbial Fermentations
Figure 9.14

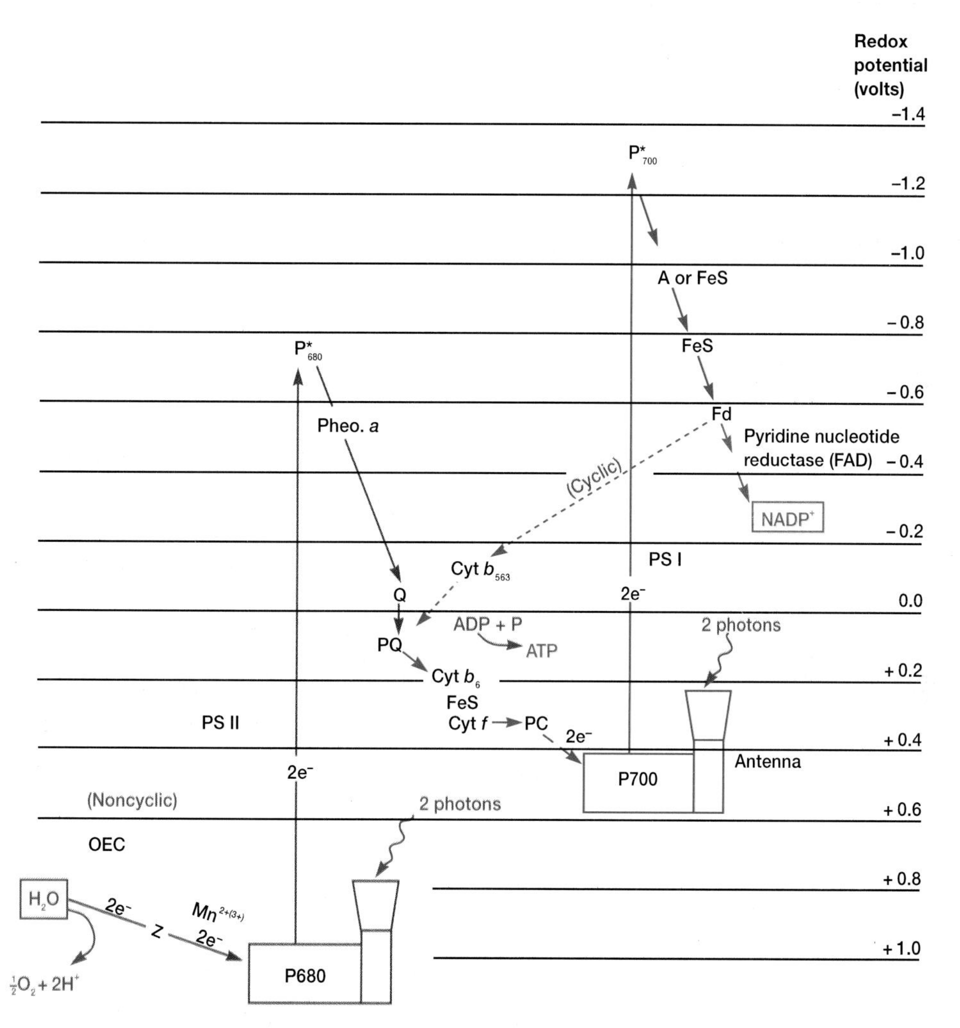

Redox potential (volts)
−1.4
−1.2
−1.0
−0.8
−0.6
−0.4
−0.2
0.0
+0.2
+0.4
+0.6
+0.8
+1.0
P*700
A or FeS
FeS
Fd
Pyridine nucleotide reductase (FAD)
NADP+
(Cyclic)
PS I
2e−
2 photons
P*680
Pheo. a
Cyt b563
Q
ADP + P
ATP
PQ
Cyt b6
FeS
Cyt f
PC
2e−
PS II
P700
Antenna
2e−
(Noncyclic)
2 photons
OEC
H2O
2e−
Z
Mn2+(3+)
2e−
P680
½O2 + 2H+

Green Plant Photosynthesis
Figure 9.25

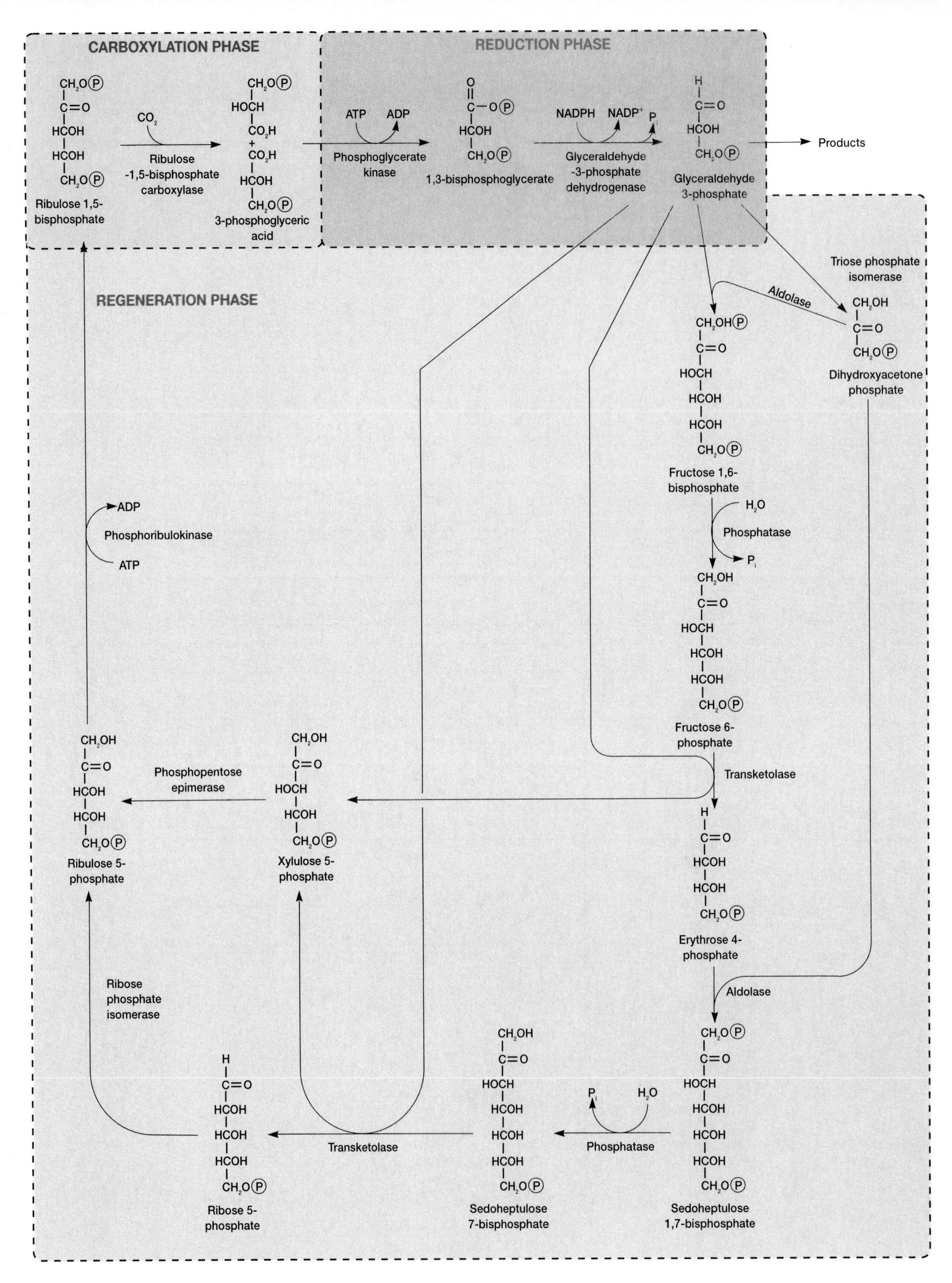
CARBOXYLATION PHASE
REDUCTION PHASE
REGENERATION PHASE
Ribulose 1,5-bisphosphate
CO2
Ribulose -1,5-bisphosphate carboxylase
3-phosphoglyceric acid
ATP
ADP
Phosphoglycerate kinase
1,3-bisphosphoglycerate
NADPH
NADP+
Pi
Glyceraldehyde -3-phosphate dehydrogenase
Glyceraldehyde 3-phosphate
Products
Triose phosphate isomerase
Aldolase
Dihydroxyacetone phosphate
Fructose 1,6-bisphosphate
H2O
Phosphatase
Pi
Fructose 6-phosphate
Transketolase
Erythrose 4-phosphate
Aldolase
Sedoheptulose 1,7-bisphosphate
Pi
H2O
Phosphatase
Sedoheptulose 7-bisphosphate
Transketolase
Ribose 5-phosphate
Ribose phosphate isomerase
Ribulose 5-phosphate
Phosphopentose epimerase
Xylulose 5-phosphate
ADP
Phosphoribulokinase
ATP

The Calvin Cycle
Figure 10.4

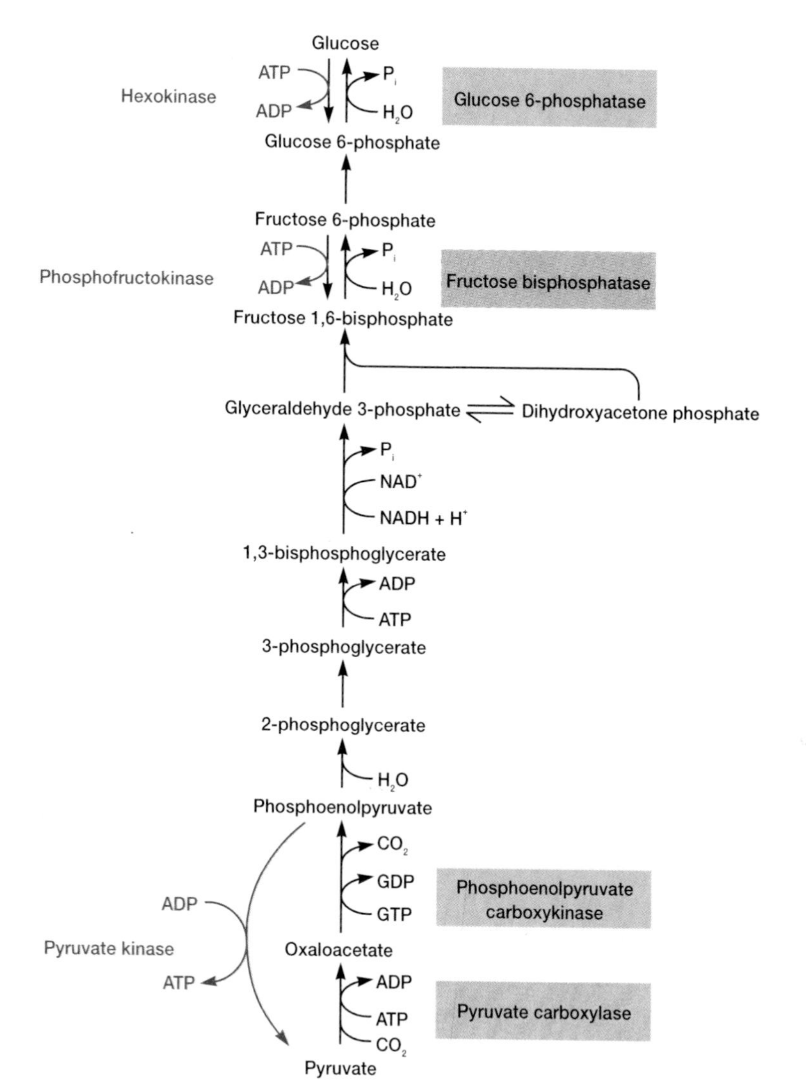
Glucose
ATP
ADP
Hexokinase
P_i
H_2O
Glucose 6-phosphatase
Glucose 6-phosphate
Fructose 6-phosphate
ATP
ADP
Phosphofructokinase
P_i
H_2O
Fructose bisphosphatase
Fructose 1,6-bisphosphate
Glyceraldehyde 3-phosphate
Dihydroxyacetone phosphate
P_i
NAD^+
$NADH + H^+$
1,3-bisphosphoglycerate
ADP
ATP
3-phosphoglycerate
2-phosphoglycerate
H_2O
Phosphoenolpyruvate
CO_2
GDP
GTP
Phosphoenolpyruvate carboxykinase
ADP
Pyruvate kinase
ATP
Oxaloacetate
ADP
ATP
CO_2
Pyruvate carboxylase
Pyruvate

Gluconeogenesis
Figure 10.5

Glutamine synthetase reaction

COOH
|
CH₂
|
CH₂   +   NH₃ + ATP ⟶
|
CH—NH₂
|
COOH

Glutamic acid

    O
    ‖
    C—NH₂
    |
    CH₂
    |
    CH₂   + ADP + Pᵢ
    |
    CH—NH₂
    |
    COOH

    Glutamine

Glutamate synthase reaction

COOH
|
C=O
|
CH₂   +
|
CH₂
|
COOH

COOH
|
CH—NH₂
|
CH₂   +
|
CH₂
|
C—NH₂
‖
O

NADPH + H⁺
or
Fd_reduced

⟶

COOH
|
CH—NH₂
|
CH₂   +
|
CH₂
|
COOH

COOH
|
CH—NH₂
|
CH₂   +
|
CH₂
|
COOH

NADP⁺
or
Fd_oxidized

α-ketoglutaric   Glutamine
acid

Glutamic acid

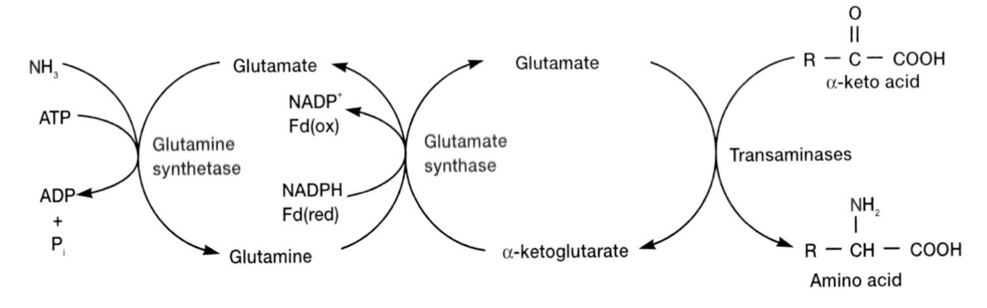

Gln Synthetase, Glu Synthase, and Ammonia Incorporation
Figure 10.11 & 10.12

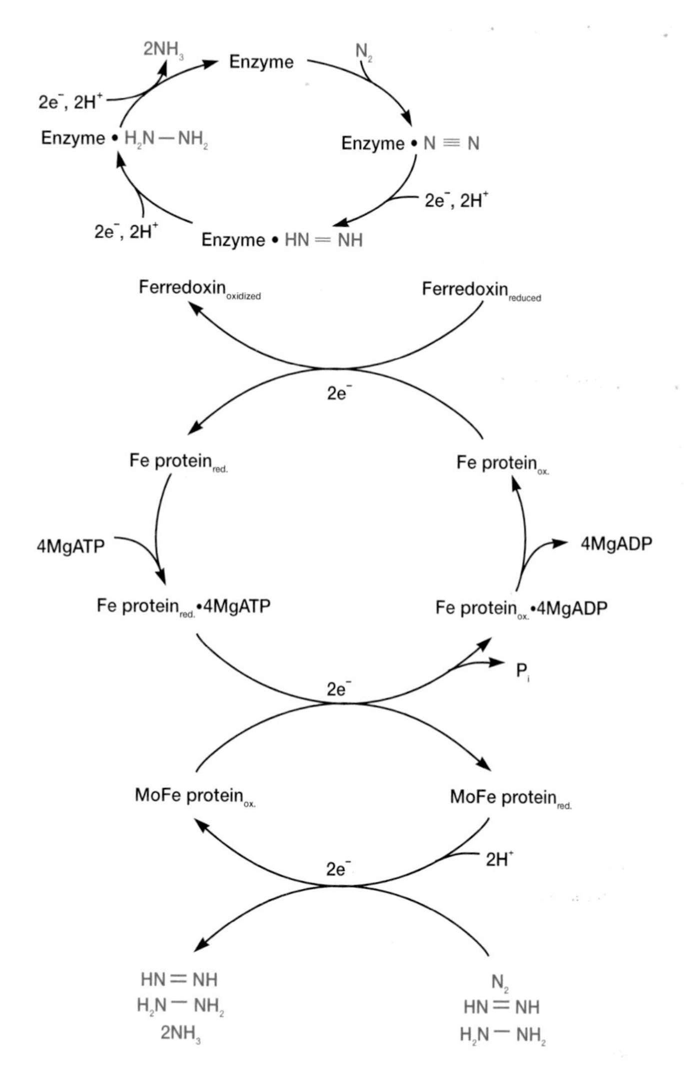

2NH3
Enzyme
N2
2e-, 2H+
Enzyme • H2N—NH2
Enzyme • N ≡ N
2e-, 2H+
2e-, 2H+
Enzyme • HN = NH
Ferredoxin oxidized
Ferredoxin reduced
2e-
Fe protein red.
Fe protein ox.
4MgATP
4MgADP
Fe protein red. •4MgATP
Fe protein ox. •4MgADP
Pi
2e-
MoFe protein ox.
MoFe protein red.
2H+
2e-
HN = NH
H2N — NH2
2NH3
N2
HN = NH
H2N — NH2

Nitrogen Fixation
Figure 10.13 & 10.15

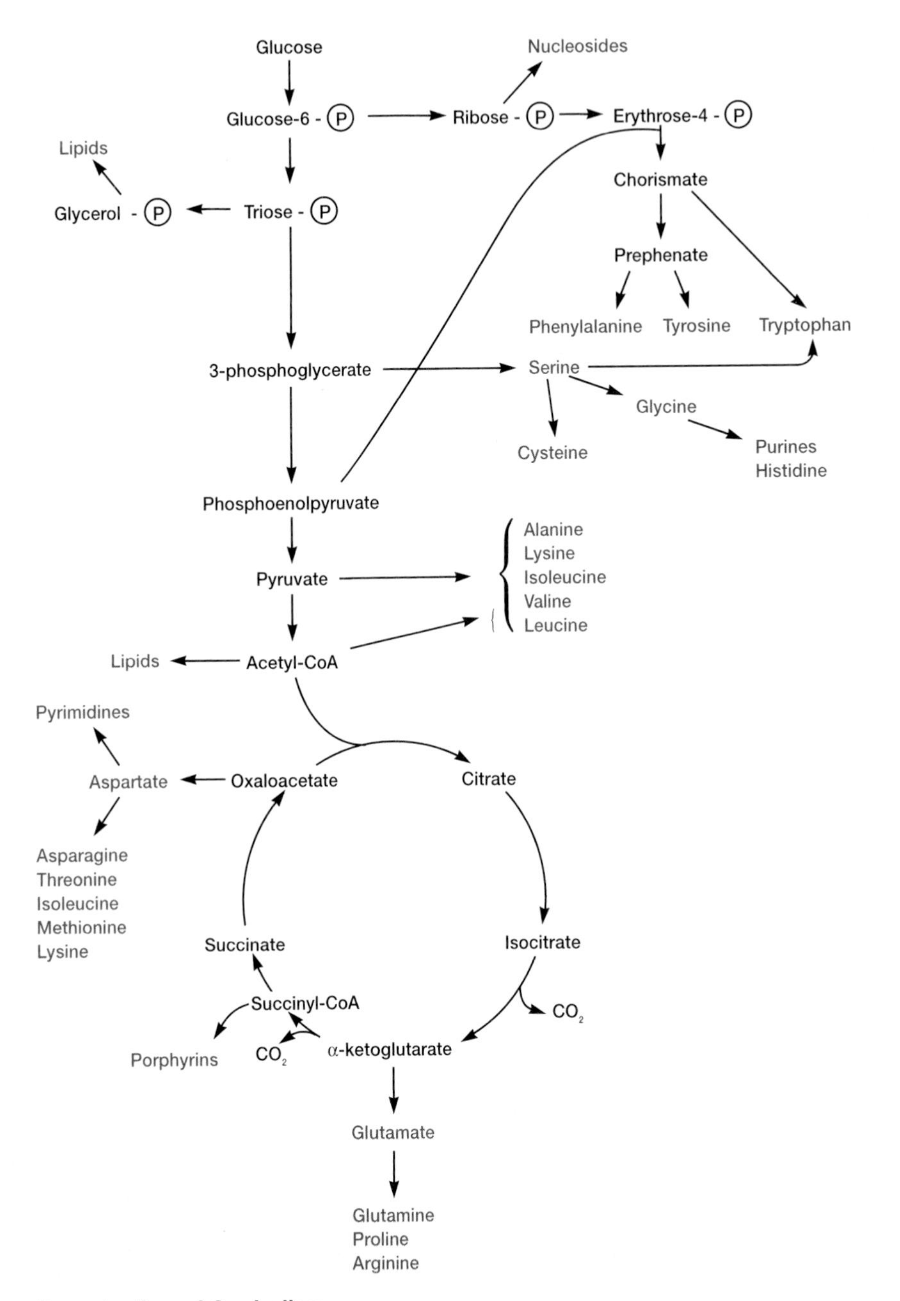

Glucose
Nucleosides
Glucose-6 - (P)
Ribose - (P)
Erythrose-4 - (P)
Lipids
Chorismate
Glycerol - (P)
Triose - (P)
Prephenate
Phenylalanine
Tyrosine
Tryptophan
3-phosphoglycerate
Serine
Glycine
Cysteine
Purines
Histidine
Phosphoenolpyruvate
Alanine
Lysine
Pyruvate
Isoleucine
Valine
Leucine
Lipids
Acetyl-CoA
Pyrimidines
Aspartate
Oxaloacetate
Citrate
Asparagine
Threonine
Isoleucine
Methionine
Lysine
Succinate
Isocitrate
Succinyl-CoA
CO_2
Porphyrins
CO_2
α-ketoglutarate
Glutamate
Glutamine
Proline
Arginine

Organization of Anabolism
Figure 10.16

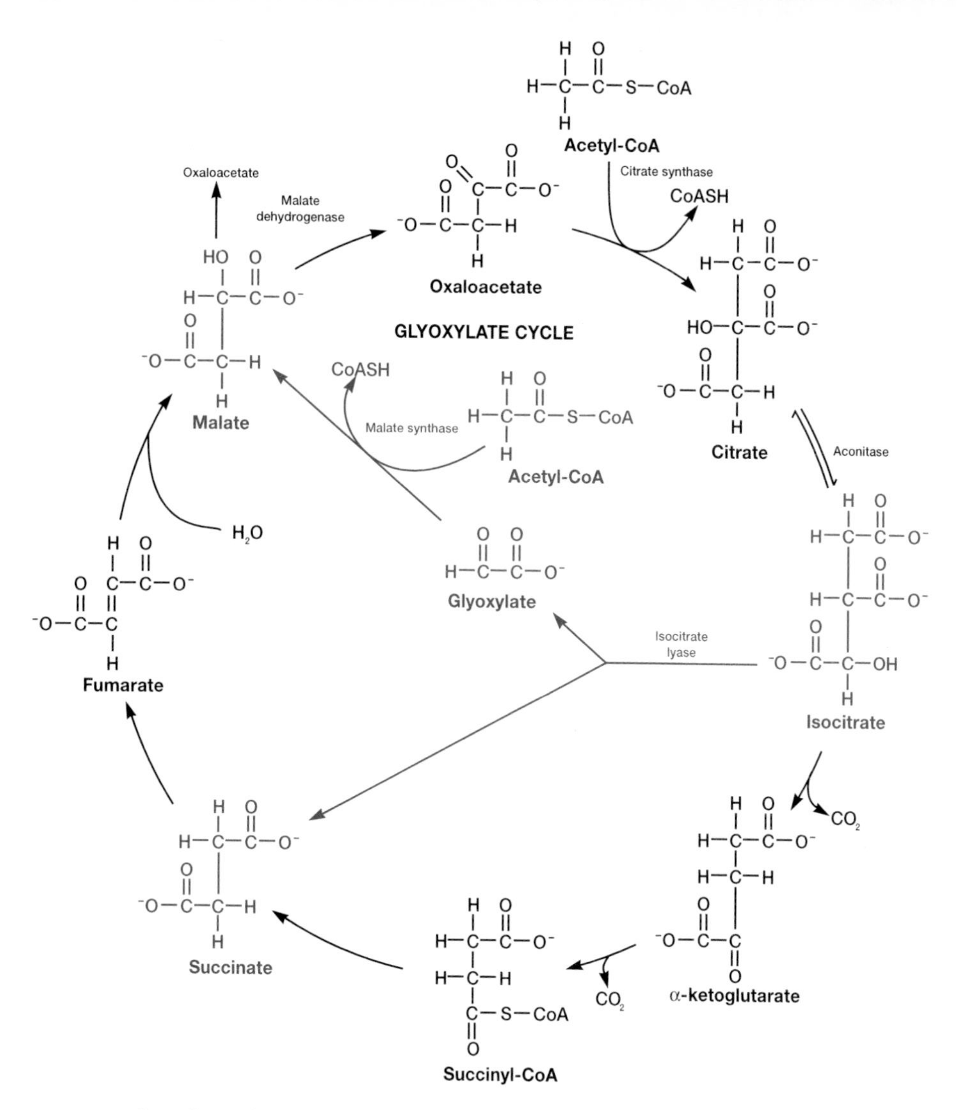

GLYOXYLATE CYCLE

Overall equation:

2 Acetyl-CoA + FAD + 2NAD$^+$ + 3H$_2$O $\longrightarrow$ Oxaloacetate + 2CoA + FADH$_2$ + 2NADH + 2H$^+$

The Glyoxylate Cycle
Figure 10.19

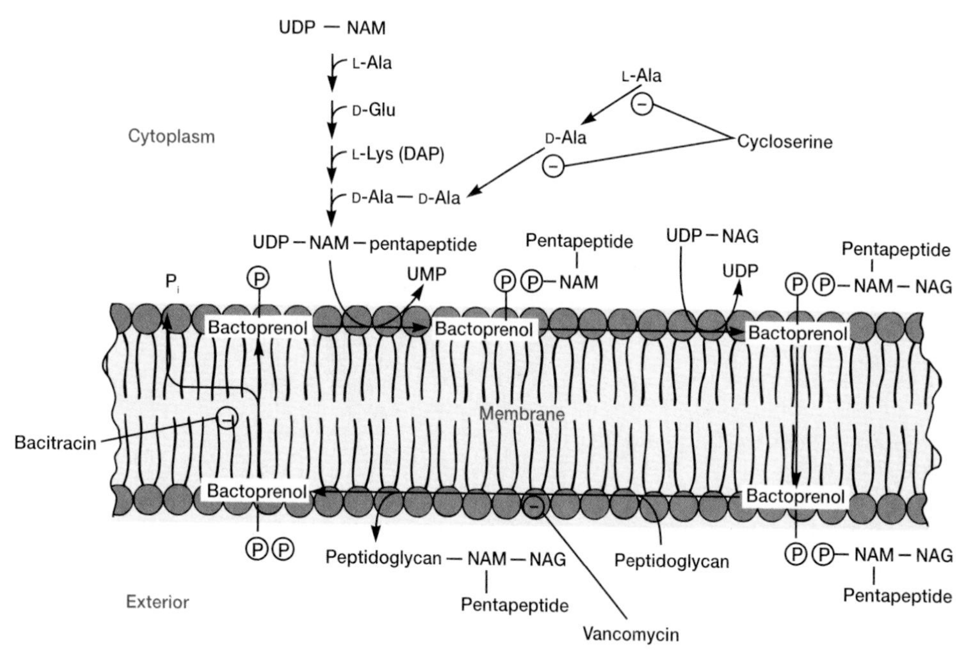

Peptidoglycan Synthesis
Figure 10.27

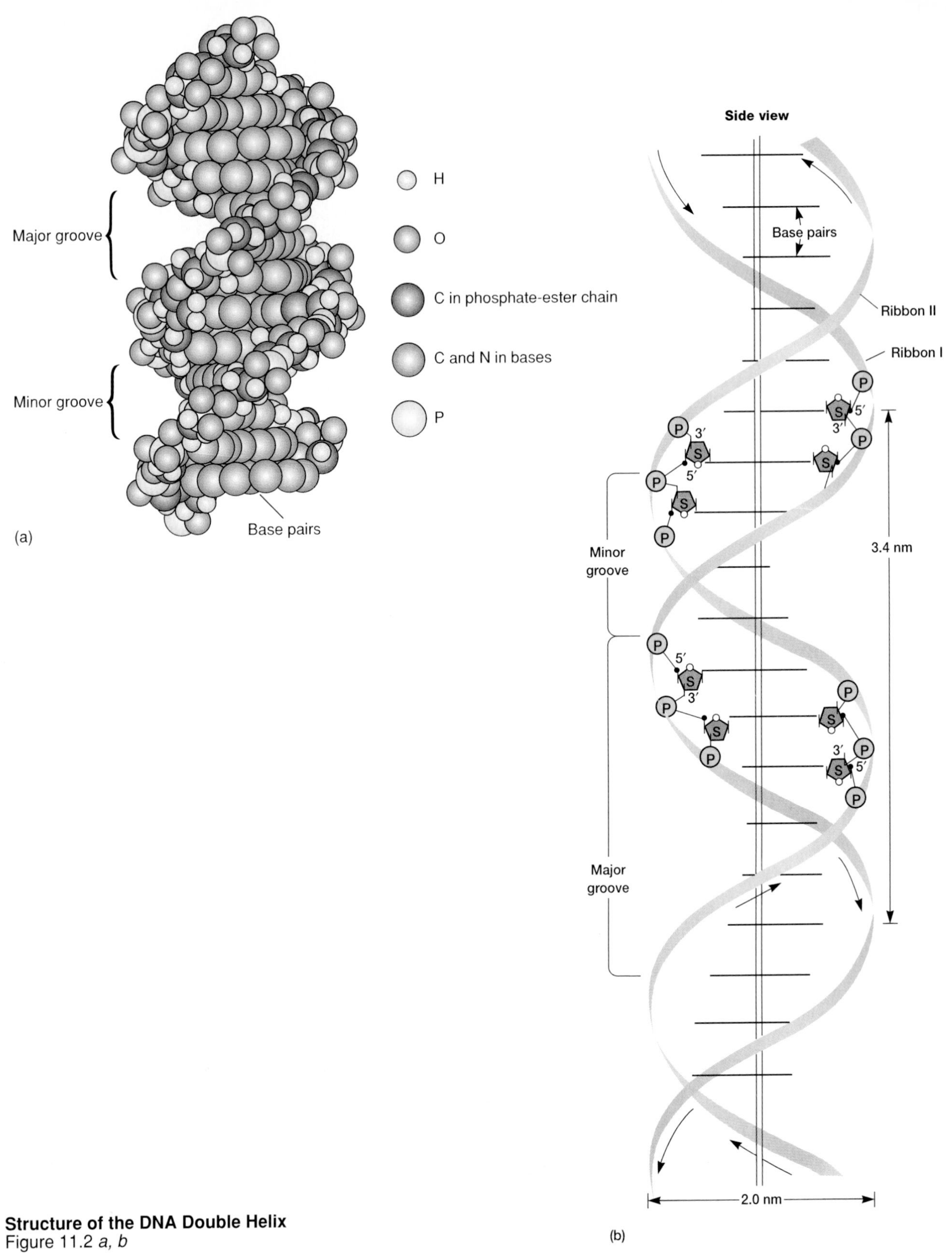

Structure of the DNA Double Helix
Figure 11.2 *a, b*

(a)

Major groove

Minor groove

Base pairs

H

O

C in phosphate-ester chain

C and N in bases

P

(b)

Side view

Base pairs

Ribbon II

Ribbon I

3.4 nm

Minor groove

Major groove

2.0 nm

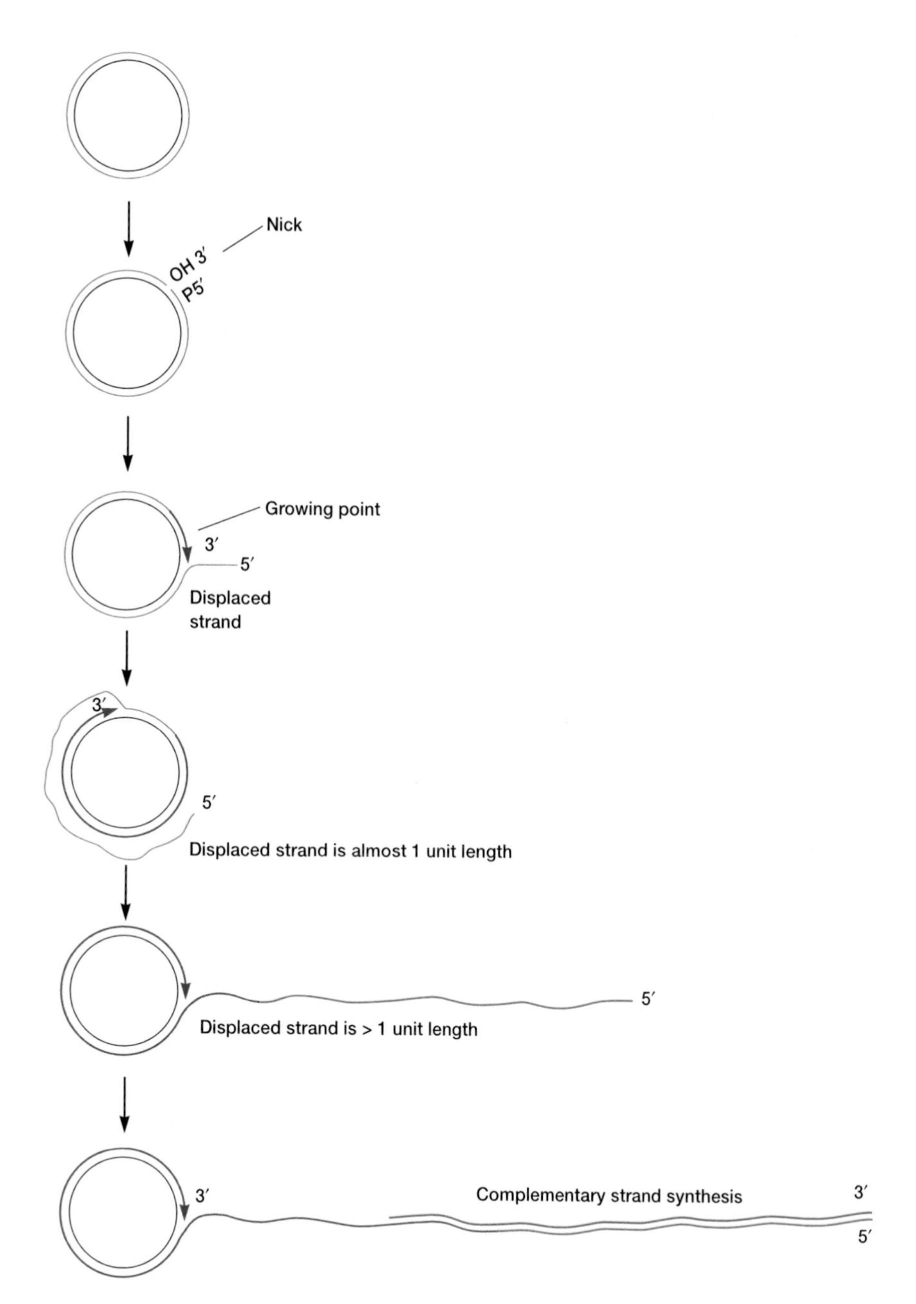

The Rolling-Circle Pattern of Replication
Figure 11.9

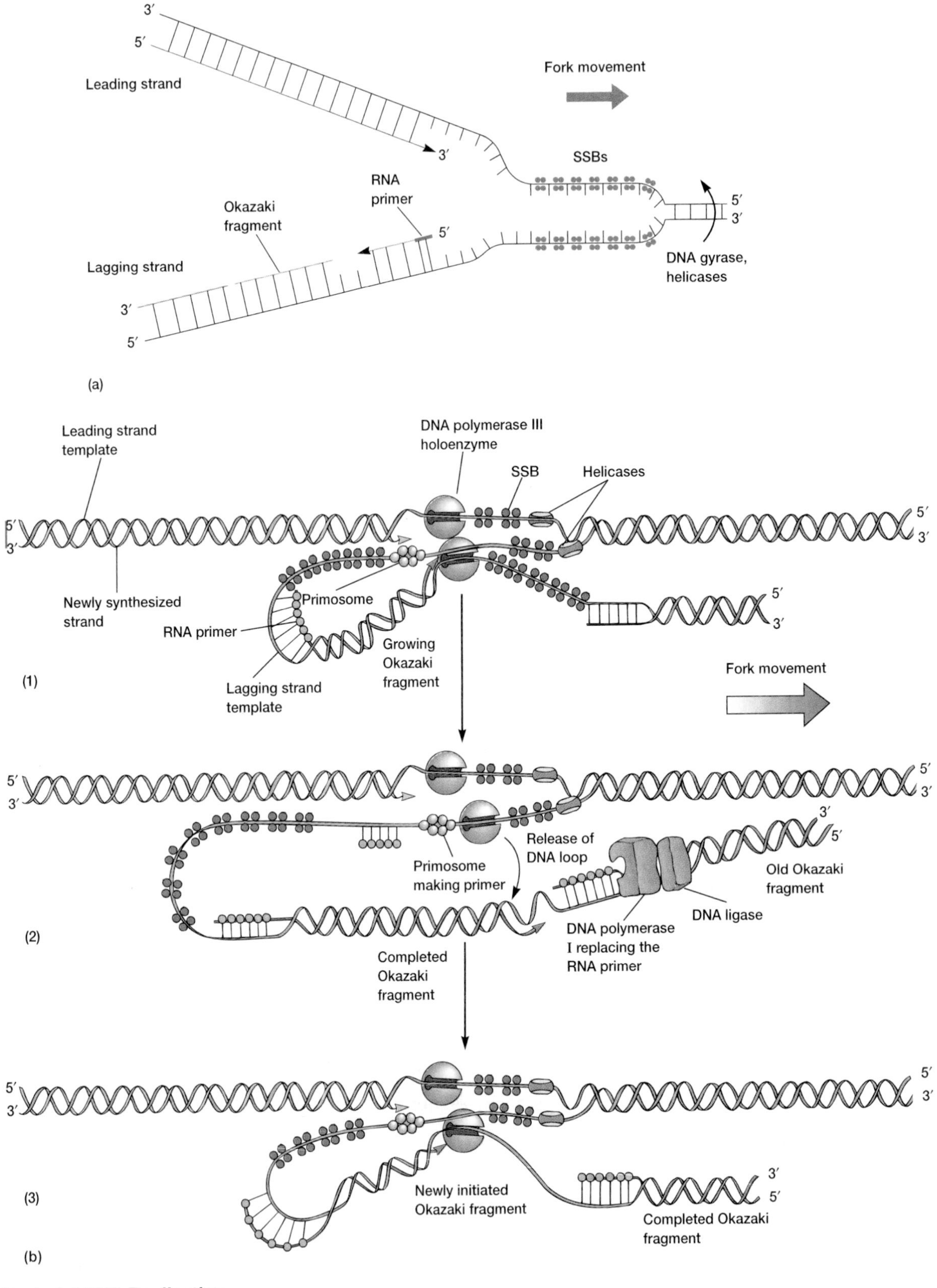
3′
5′
Leading strand
3′
Fork movement
SSBs
RNA
primer
Okazaki
fragment
5′
5′
3′
Lagging strand
3′
5′
DNA gyrase,
helicases
(a)
Leading strand
template
DNA polymerase III
holoenzyme
SSB
Helicases
5′
3′
5′
3′
Newly synthesized
strand
Primosome
RNA primer
5′
3′
Growing
Okazaki
fragment
(1)
Lagging strand
template
Fork movement
5′
3′
5′
3′
3′
5′
Release of
DNA loop
Primosome
making primer
Old Okazaki
fragment
DNA ligase
DNA polymerase
I replacing the
RNA primer
(2)
Completed
Okazaki
fragment
5′
3′
5′
3′
3′
5′
(3)
Newly initiated
Okazaki fragment
Completed Okazaki
fragment
(b)

Bacterial DNA Replication
Figure 11.12 *a, b*

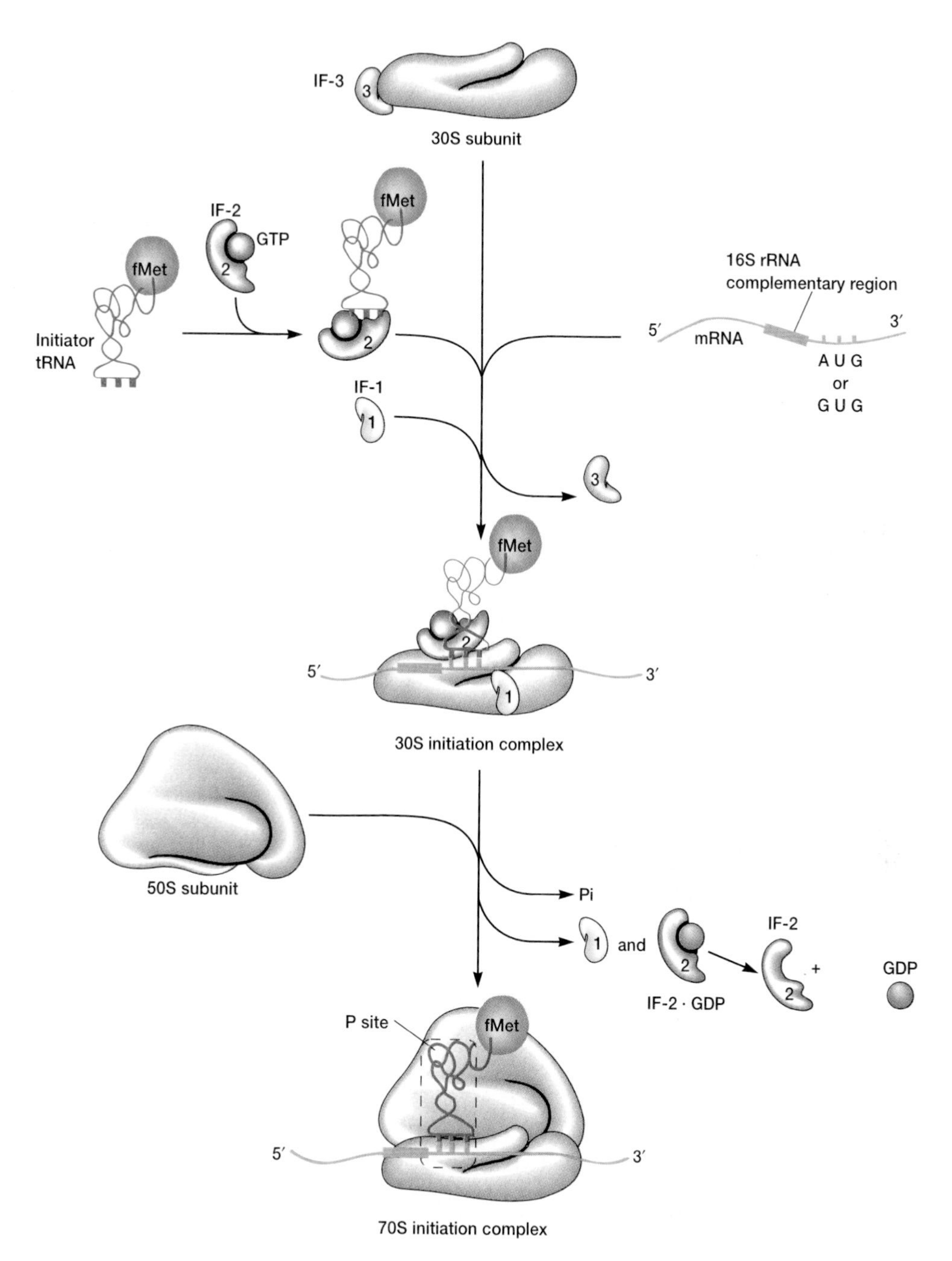

IF-3

30S subunit

IF-2

GTP

2

fMet

fMet

2

Initiator
tRNA

IF-1

1

16S rRNA
complementary region

5′

mRNA

3′

A U G
or
G U G

3

fMet

2

5′

3′

1

30S initiation complex

50S subunit

Pi

1

and

2

IF-2

2

+

GDP

IF-2 · GDP

P site

fMet

5′

3′

70S initiation complex

Initiation of Protein Synthesis
Figure 11.28

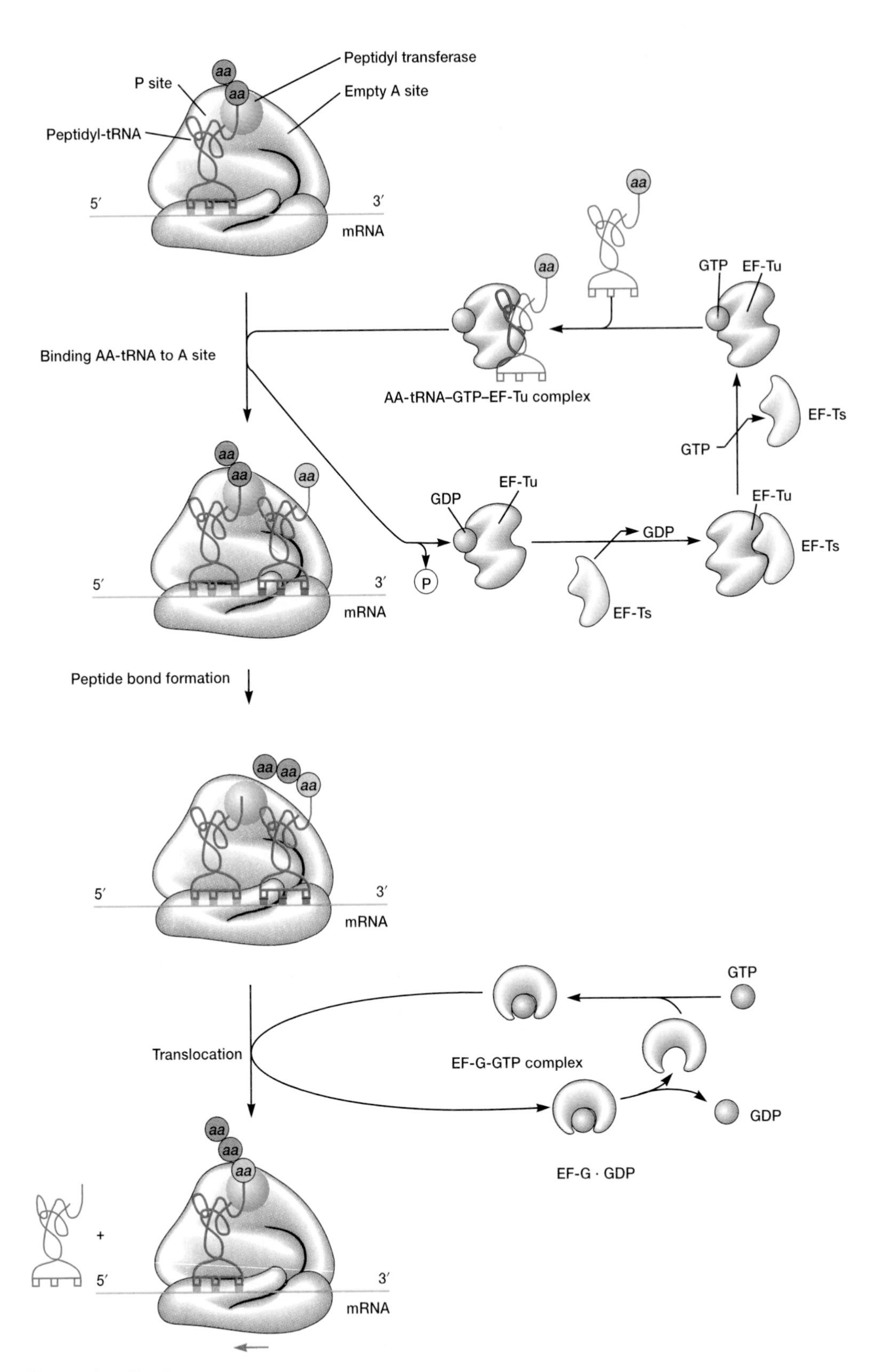

Peptidyl transferase
P site
Empty A site
Peptidyl-tRNA
aa
5′
3′
mRNA
Binding AA-tRNA to A site
AA-tRNA–GTP–EF-Tu complex
GTP
EF-Tu
EF-Ts
GDP
P
Peptide bond formation
Translocation
EF-G-GTP complex
EF-G · GDP
+

Elongation Cycle
Figure 11.29

Genes transcribed

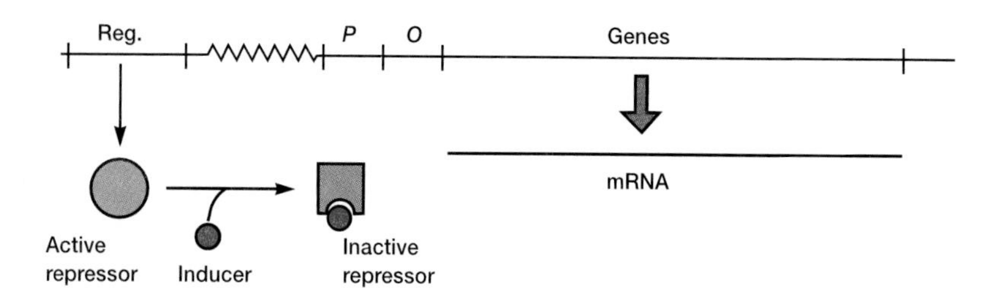

Genes not transcribed

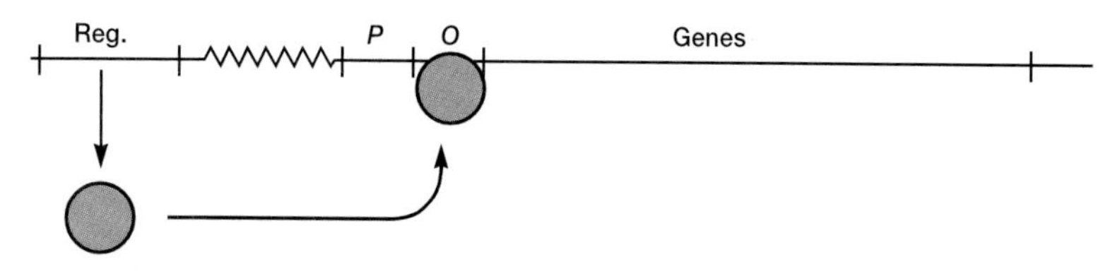

Gene Induction
Figure 12.10

Genes not transcribed

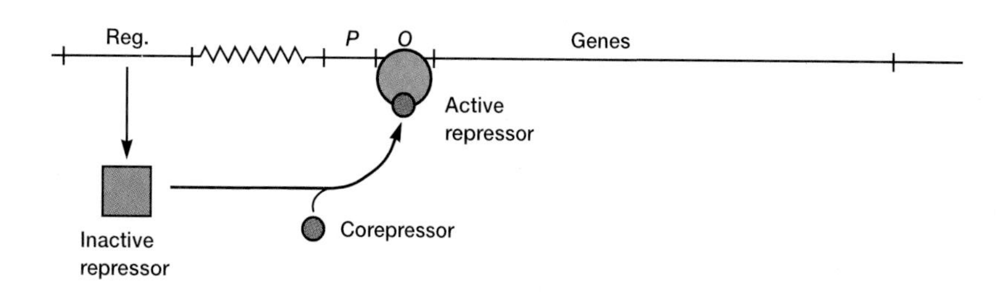

Genes transcribed

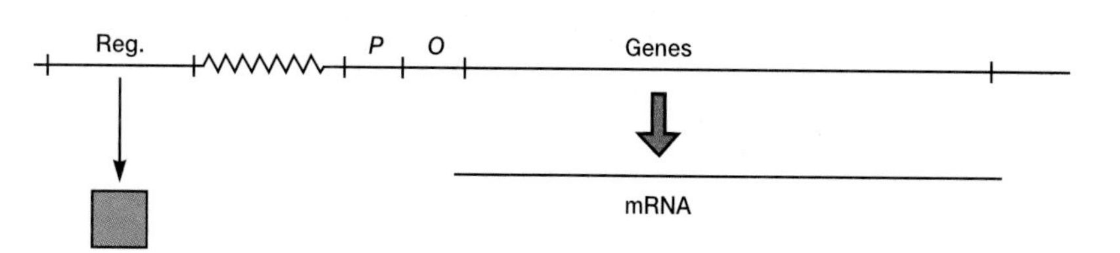

Gene Repression
Figure 12.11

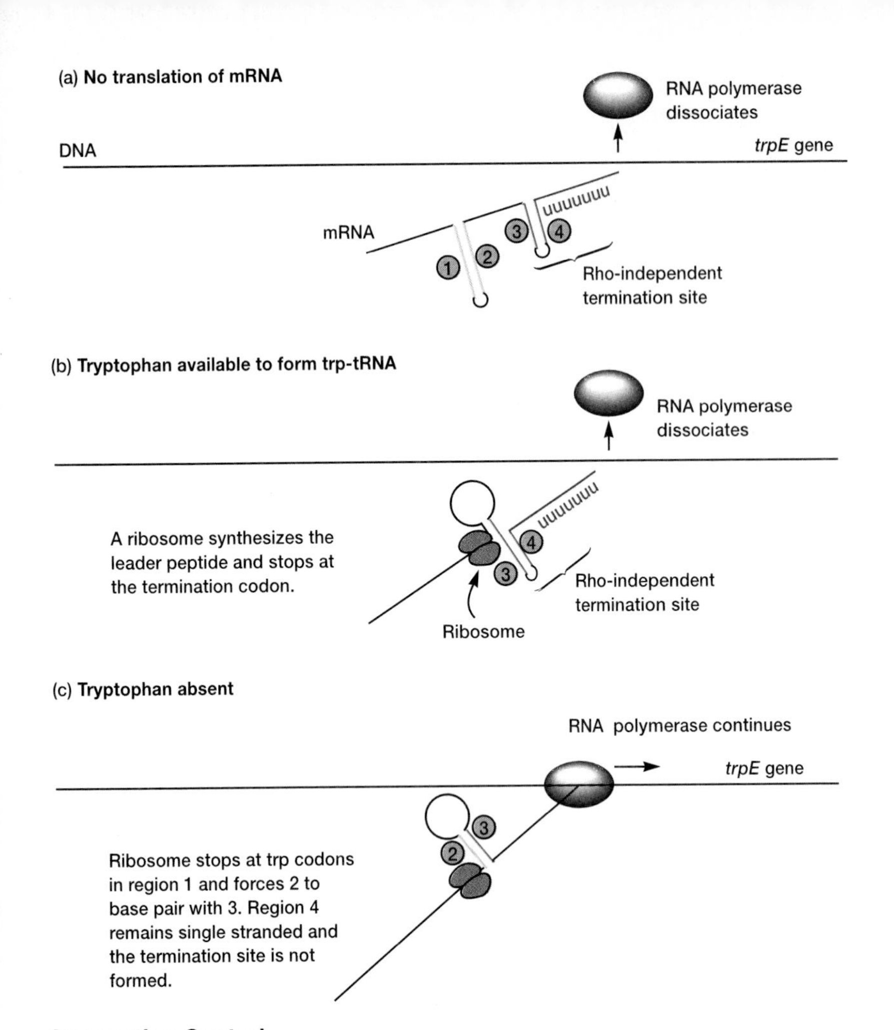

(a) **No translation of mRNA**

DNA

RNA polymerase dissociates

trpE gene

mRNA

uuuuuuu

Rho-independent termination site

(b) **Tryptophan available to form trp-tRNA**

RNA polymerase dissociates

A ribosome synthesizes the leader peptide and stops at the termination codon.

uuuuuuu

Rho-independent termination site

Ribosome

(c) **Tryptophan absent**

RNA polymerase continues

trpE gene

Ribosome stops at trp codons in region 1 and forces 2 to base pair with 3. Region 4 remains single stranded and the termination site is not formed.

Attenuation Control
Figure 12.17

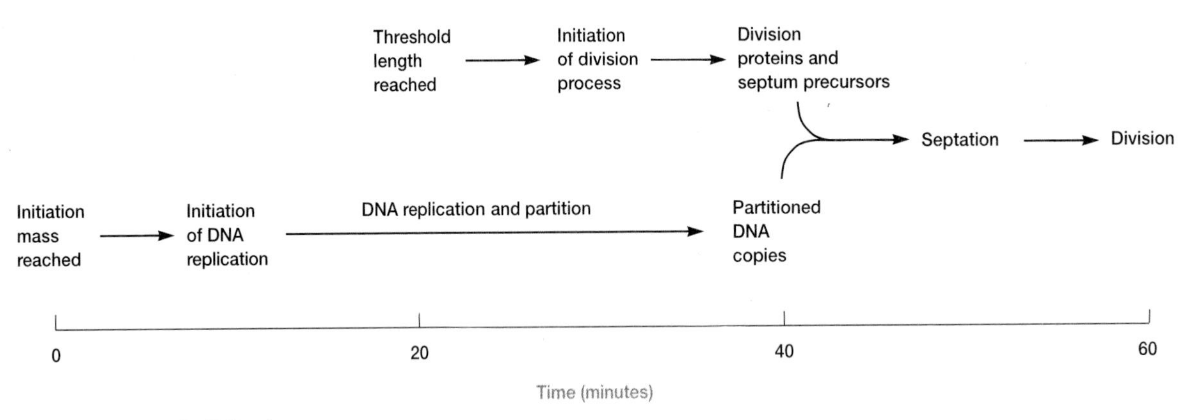

Threshold length reached → Initiation of division process → Division proteins and septum precursors

Septation → Division

Initiation mass reached → Initiation of DNA replication → DNA replication and partition → Partitioned DNA copies

Time (minutes)

0   20   40   60

Control of the Cell Cycle
Figure 12.18

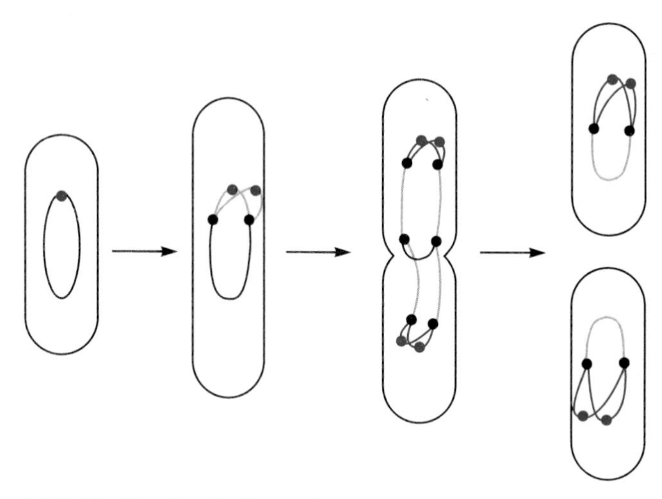

DNA Replication in Bacteria
Figure 12.19

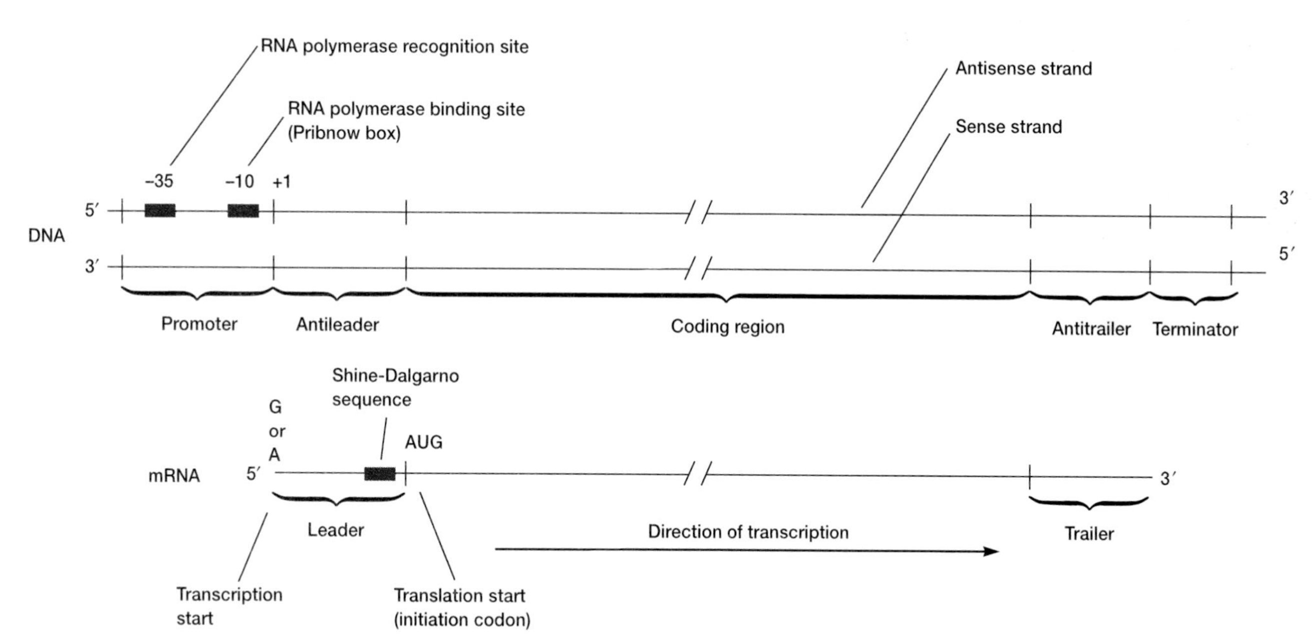

A Bacterial Structural Gene
Figure 13.6

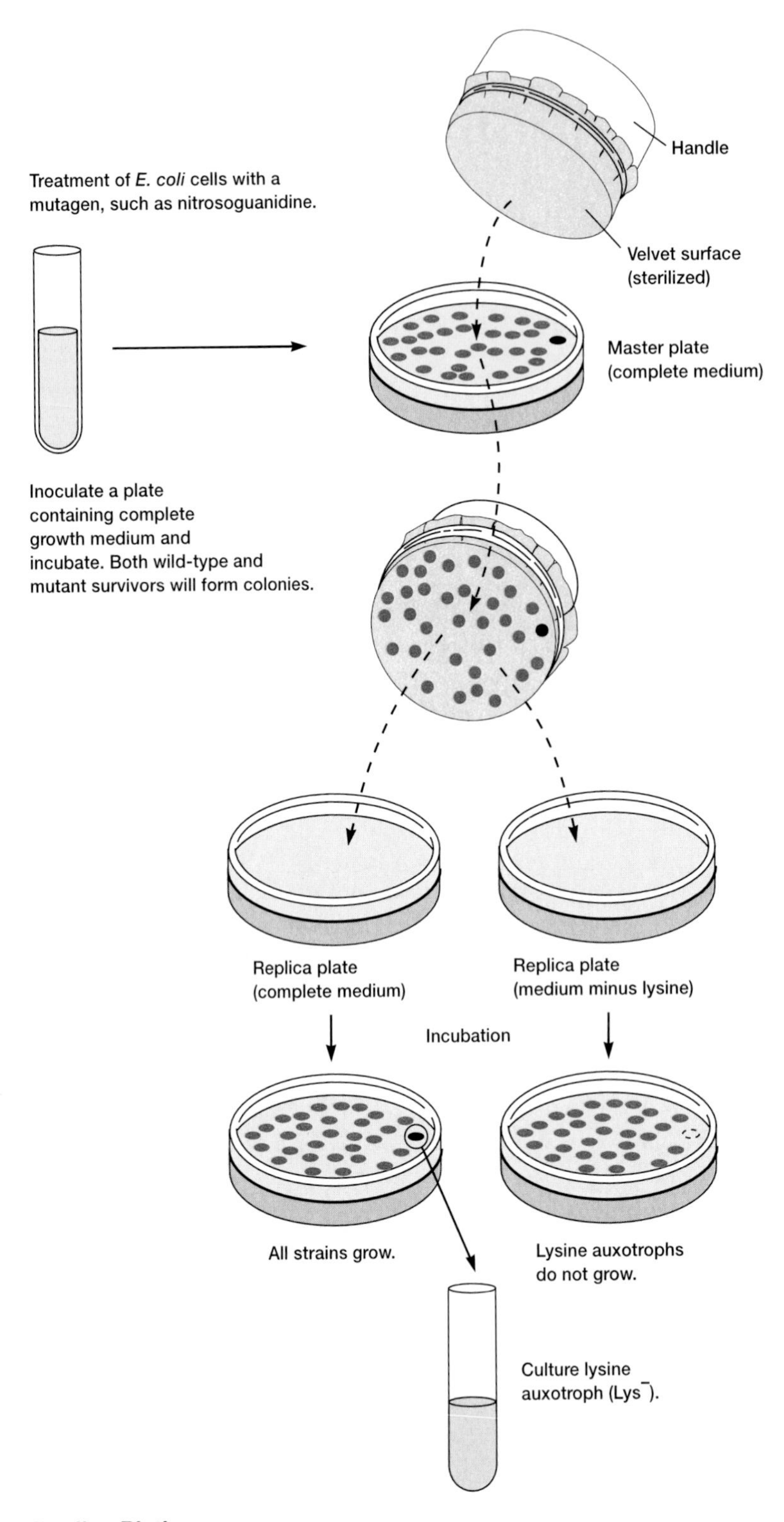

Treatment of *E. coli* cells with a mutagen, such as nitrosoguanidine.

Inoculate a plate containing complete growth medium and incubate. Both wild-type and mutant survivors will form colonies.

Handle

Velvet surface (sterilized)

Master plate (complete medium)

Replica plate (complete medium)

Replica plate (medium minus lysine)

Incubation

All strains grow.

Lysine auxotrophs do not grow.

Culture lysine auxotroph (Lys⁻).

Replica Plating
Figure 13.15

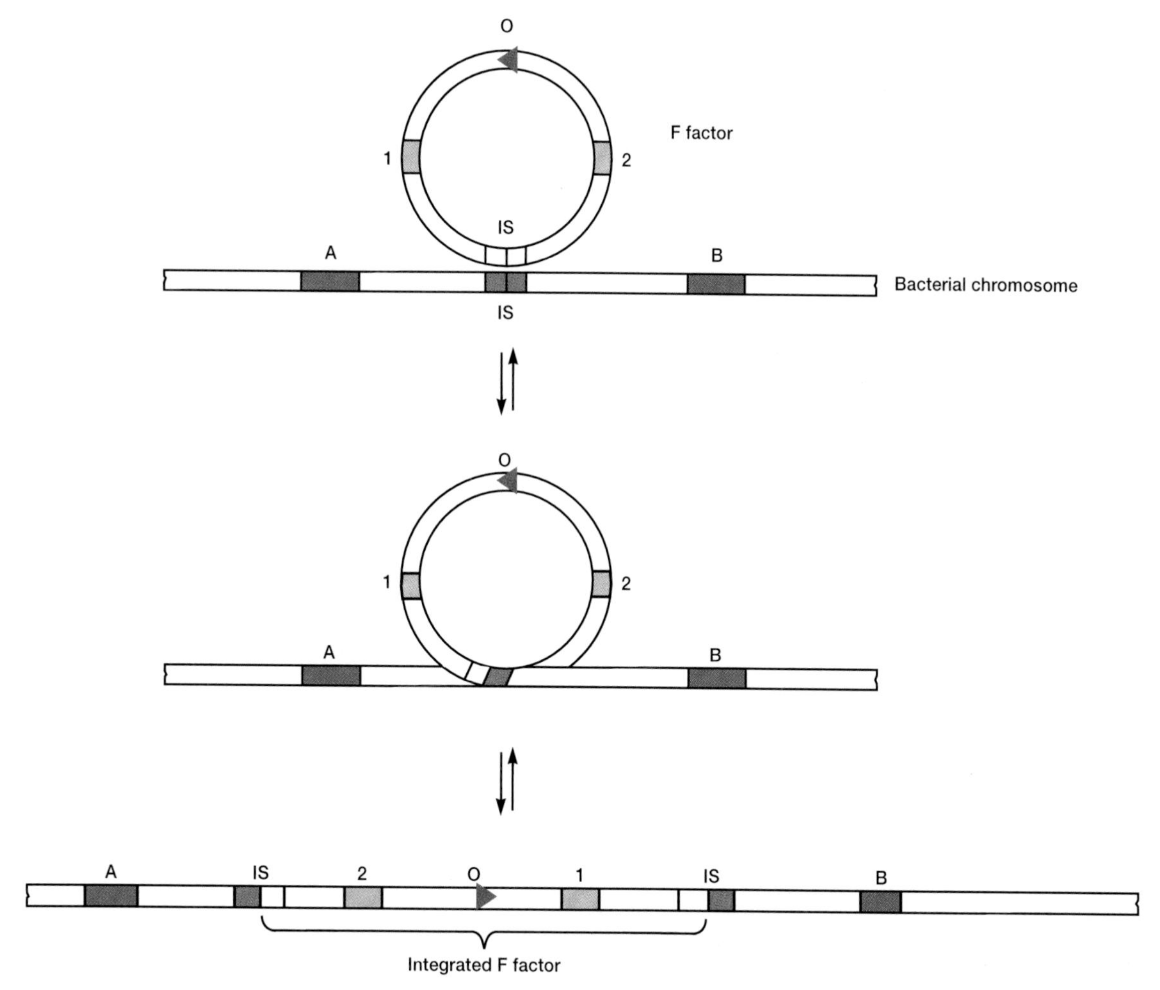

O

1   2

F factor

IS

A   B   Bacterial chromosome

IS

O

1   2

A   B

A   IS   2   O   1   IS   B

Integrated F factor

F Plasmid Integration
Figure 14.6

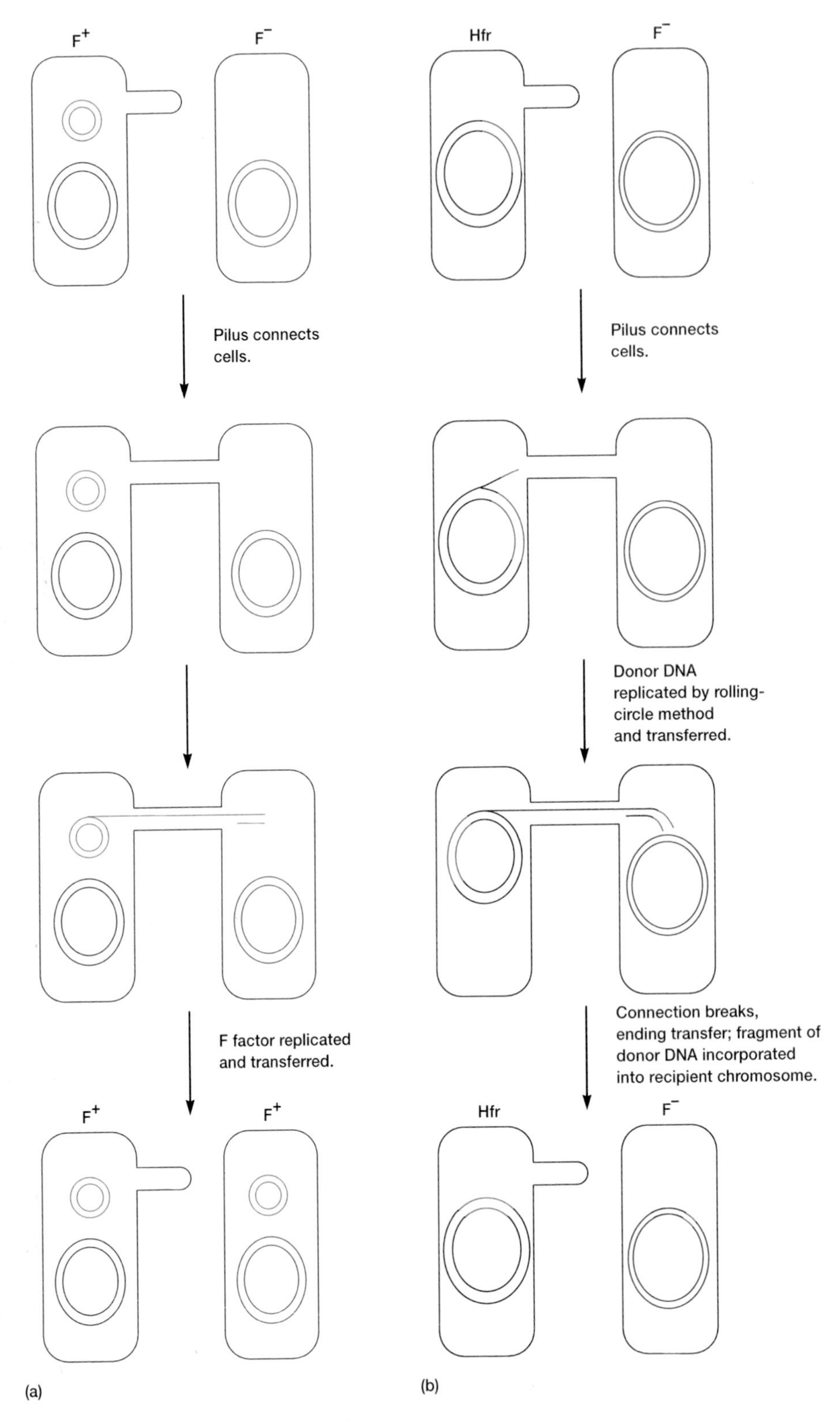

F⁺   F⁻

Hfr   F⁻

Pilus connects
cells.

Pilus connects
cells.

Donor DNA
replicated by rolling-
circle method
and transferred.

F factor replicated
and transferred.

Connection breaks,
ending transfer; fragment of
donor DNA incorporated
into recipient chromosome.

F⁺   F⁺

Hfr   F⁻

(a)

(b)

Mechanism of Bacterial Conjugation
Figure 14.13

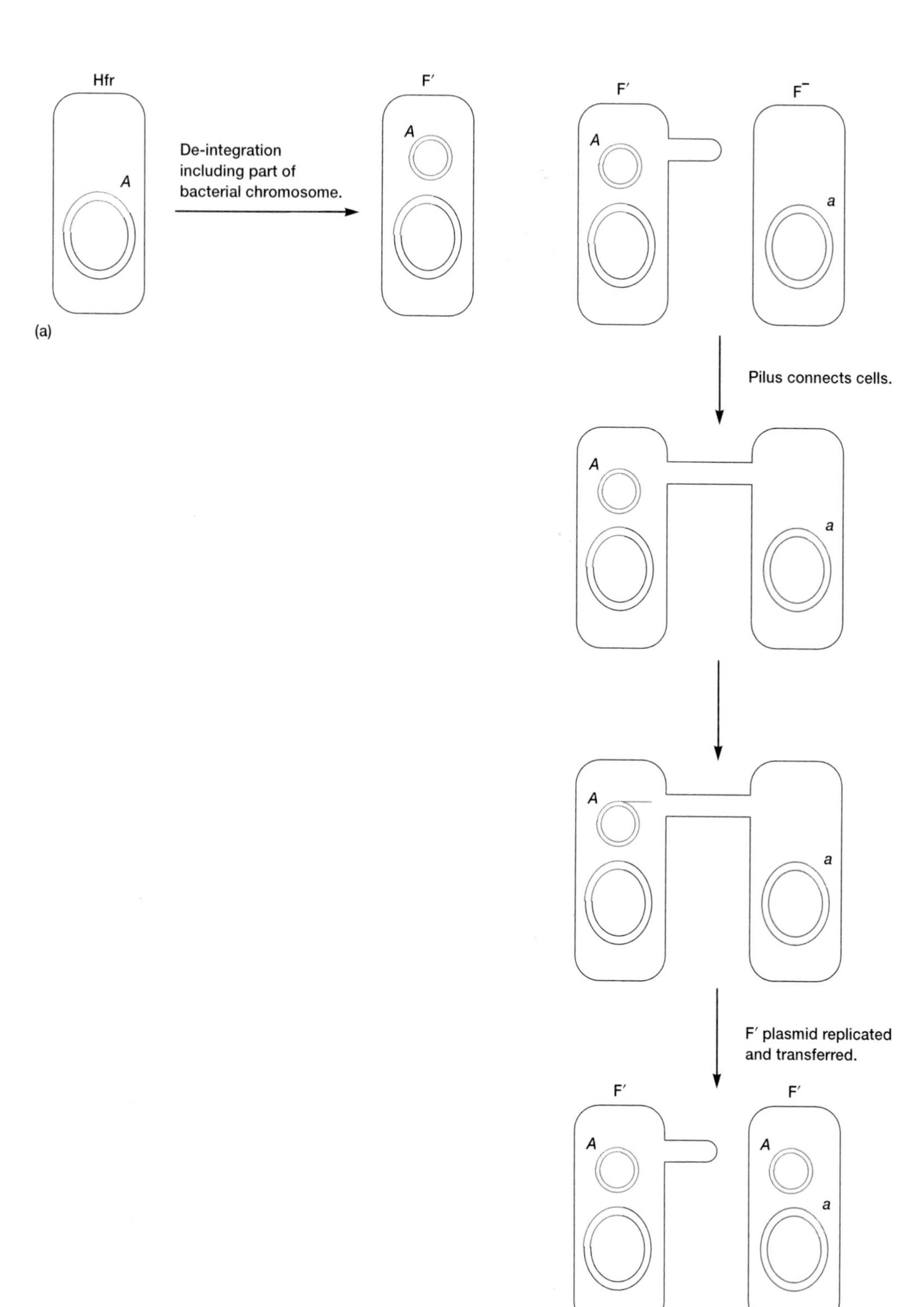

Hfr

De-integration
including part of
bacterial chromosome.

F′

(a)

F′   F⁻

Pilus connects cells.

F′ plasmid replicated
and transferred.

F′   F′

(b)

F′ Conjugation
Figure 14.14

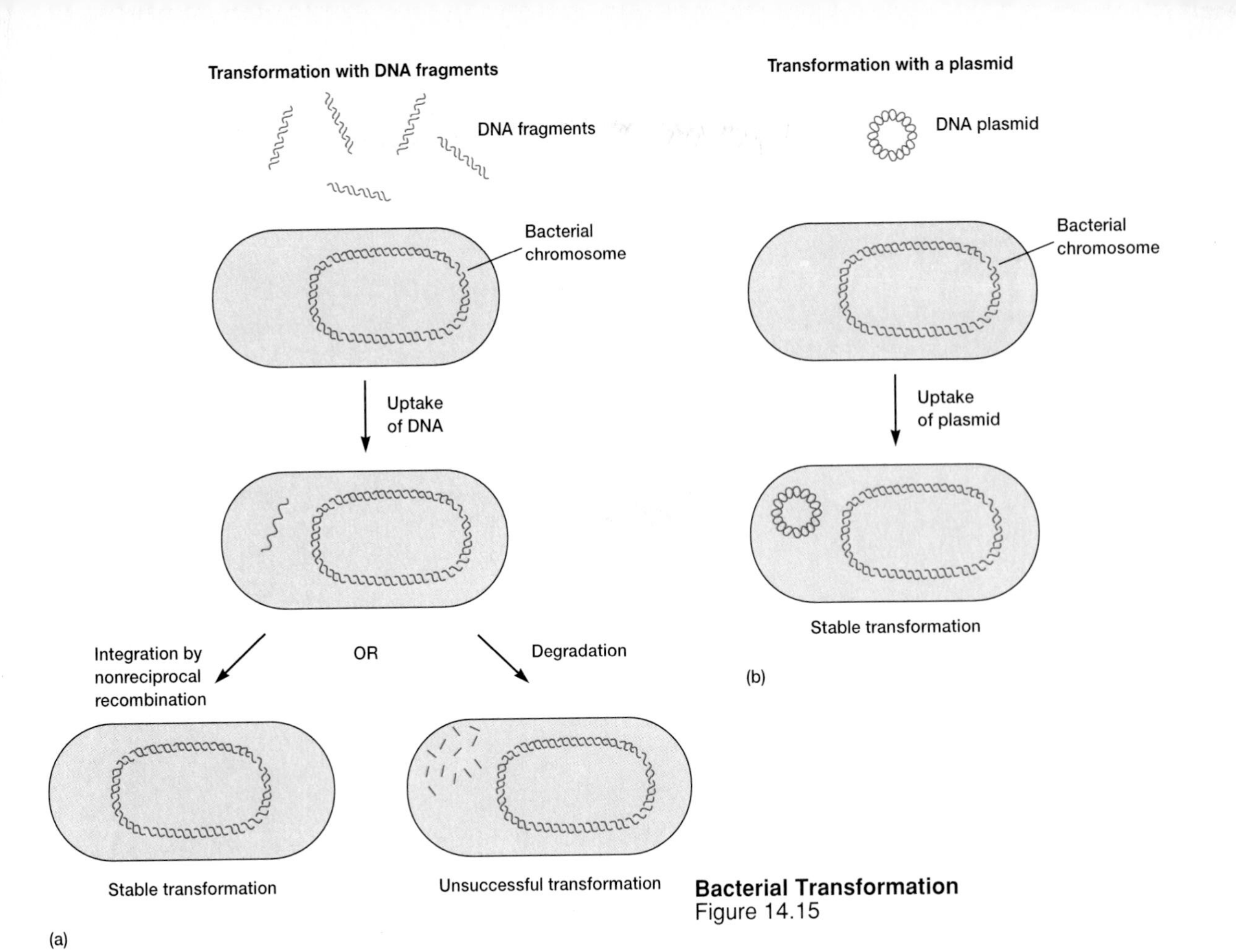

Transformation with DNA fragments

DNA fragments

Bacterial
chromosome

Uptake
of DNA

Integration by
nonreciprocal
recombination

OR

Degradation

Stable transformation

Unsuccessful transformation

(a)

Transformation with a plasmid

DNA plasmid

Bacterial
chromosome

Uptake
of plasmid

Stable transformation

(b)

Bacterial Transformation
Figure 14.15

1

2

4

3

Nucleotides

Mechanism of Transformation
Figure 14.16

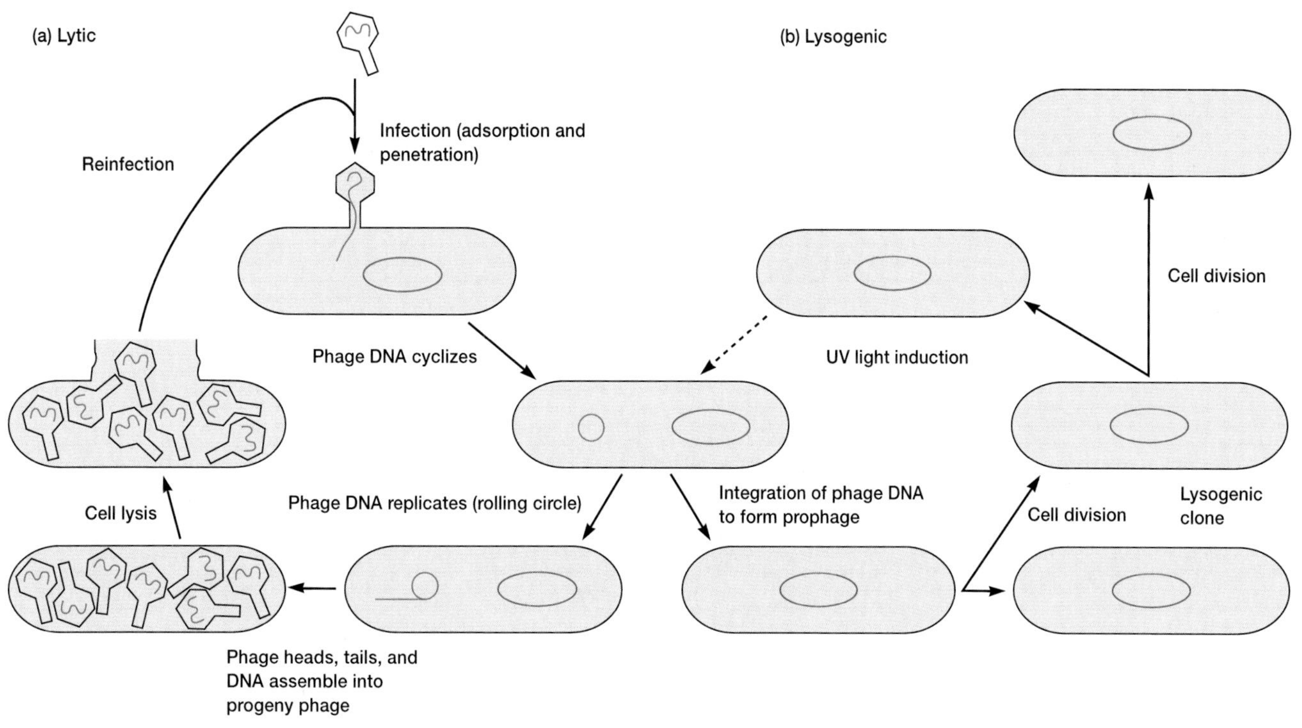

(a) Lytic

(b) Lysogenic

Reinfection

Infection (adsorption and penetration)

Phage DNA cyclizes

Cell lysis

Phage DNA replicates (rolling circle)

Phage heads, tails, and DNA assemble into progeny phage

Integration of phage DNA to form prophage

UV light induction

Cell division

Cell division

Lysogenic clone

Lytic versus Lysogenic Infection by Phage Lambda
Figure 14.17

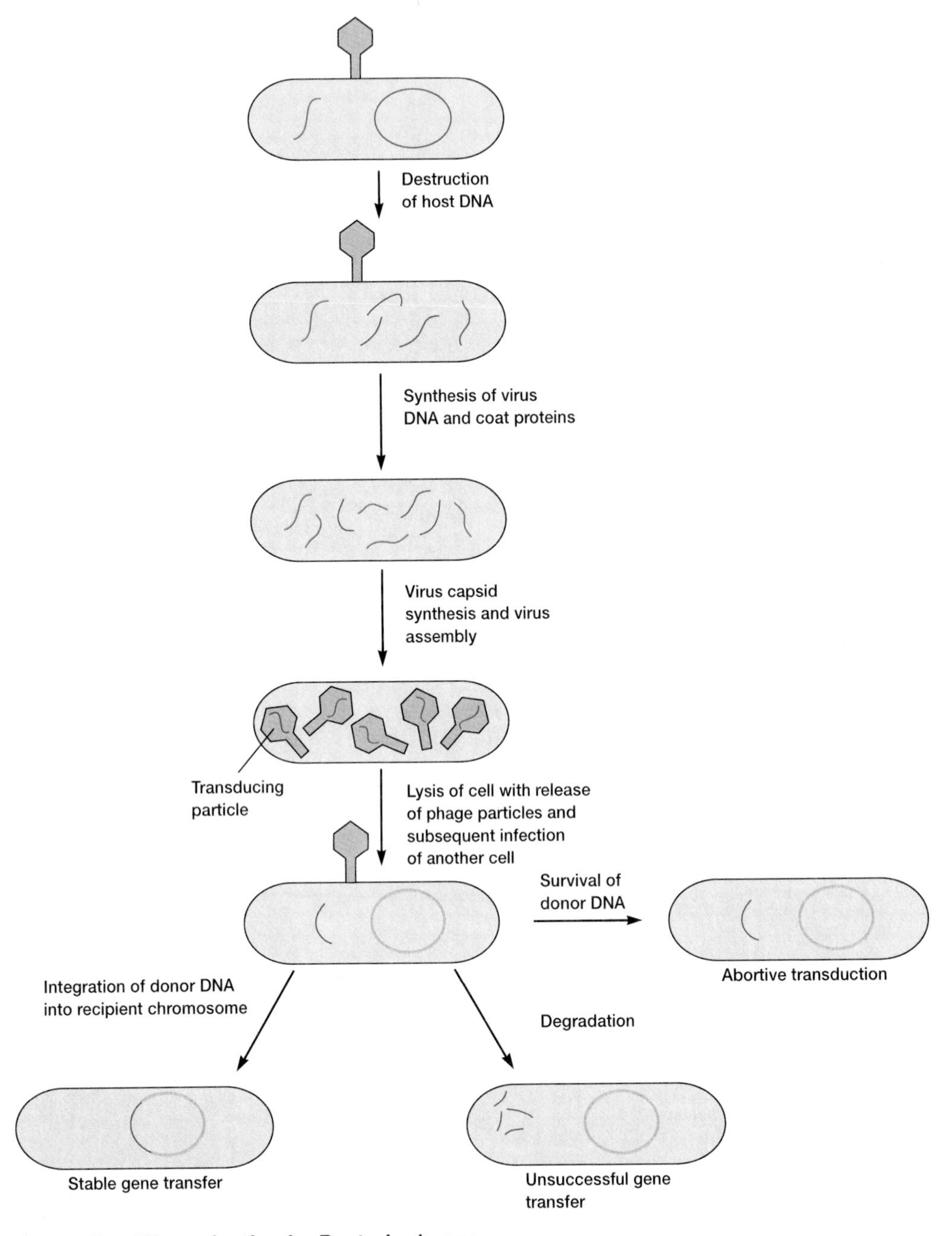

Destruction of host DNA

Synthesis of virus DNA and coat proteins

Virus capsid synthesis and virus assembly

Transducing particle

Lysis of cell with release of phage particles and subsequent infection of another cell

Survival of donor DNA

Abortive transduction

Integration of donor DNA into recipient chromosome

Degradation

Stable gene transfer

Unsuccessful gene transfer

Generalized Transduction by Bacteriophages
Figure 14.18

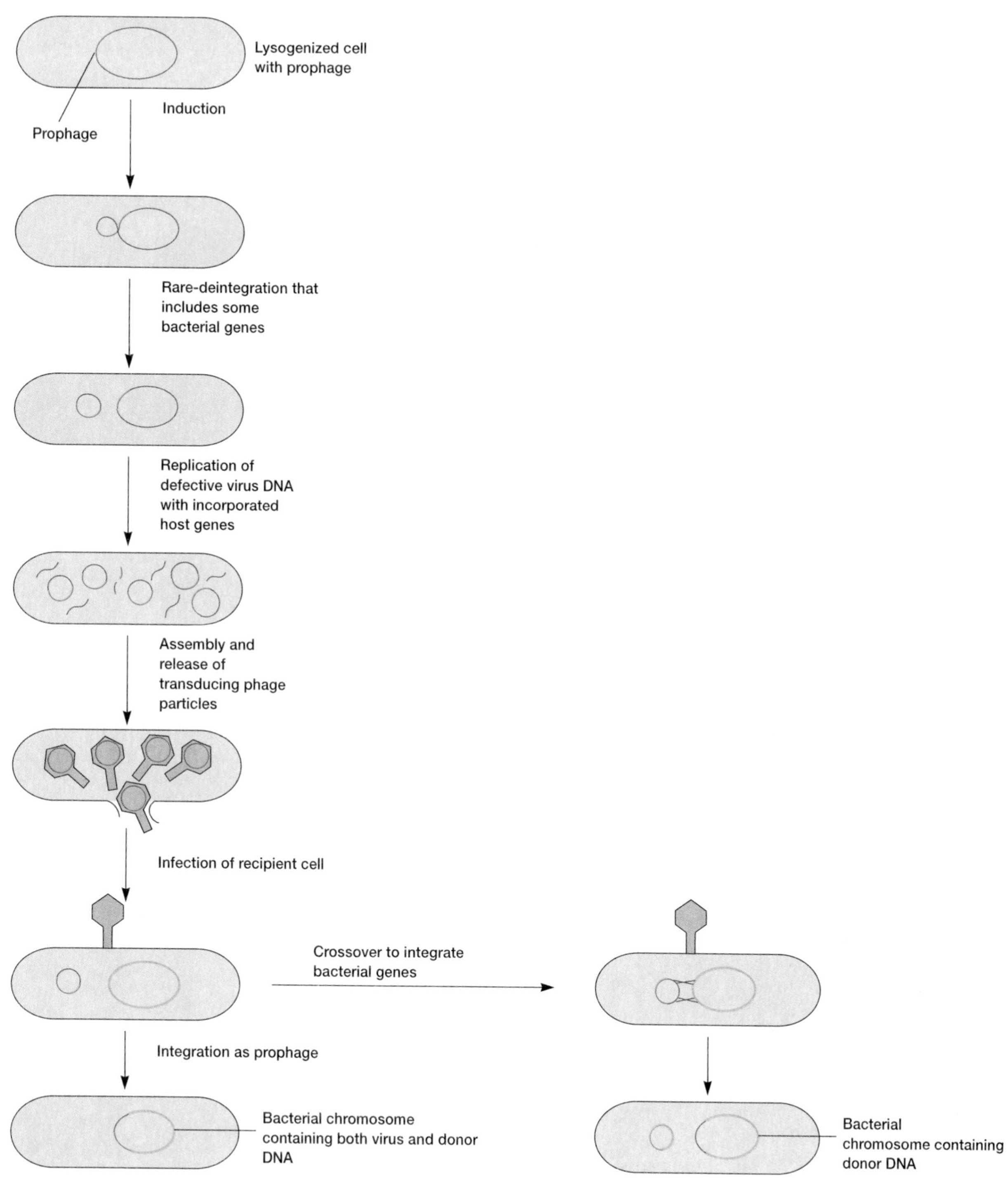

Lysogenized cell with prophage

Prophage

Induction

Rare-deintegration that includes some bacterial genes

Replication of defective virus DNA with incorporated host genes

Assembly and release of transducing phage particles

Infection of recipient cell

Crossover to integrate bacterial genes

Integration as prophage

Bacterial chromosome containing both virus and donor DNA

Bacterial chromosome containing donor DNA

Specialized Transduction by a Temperate Bacteriophage
Figure 14.19

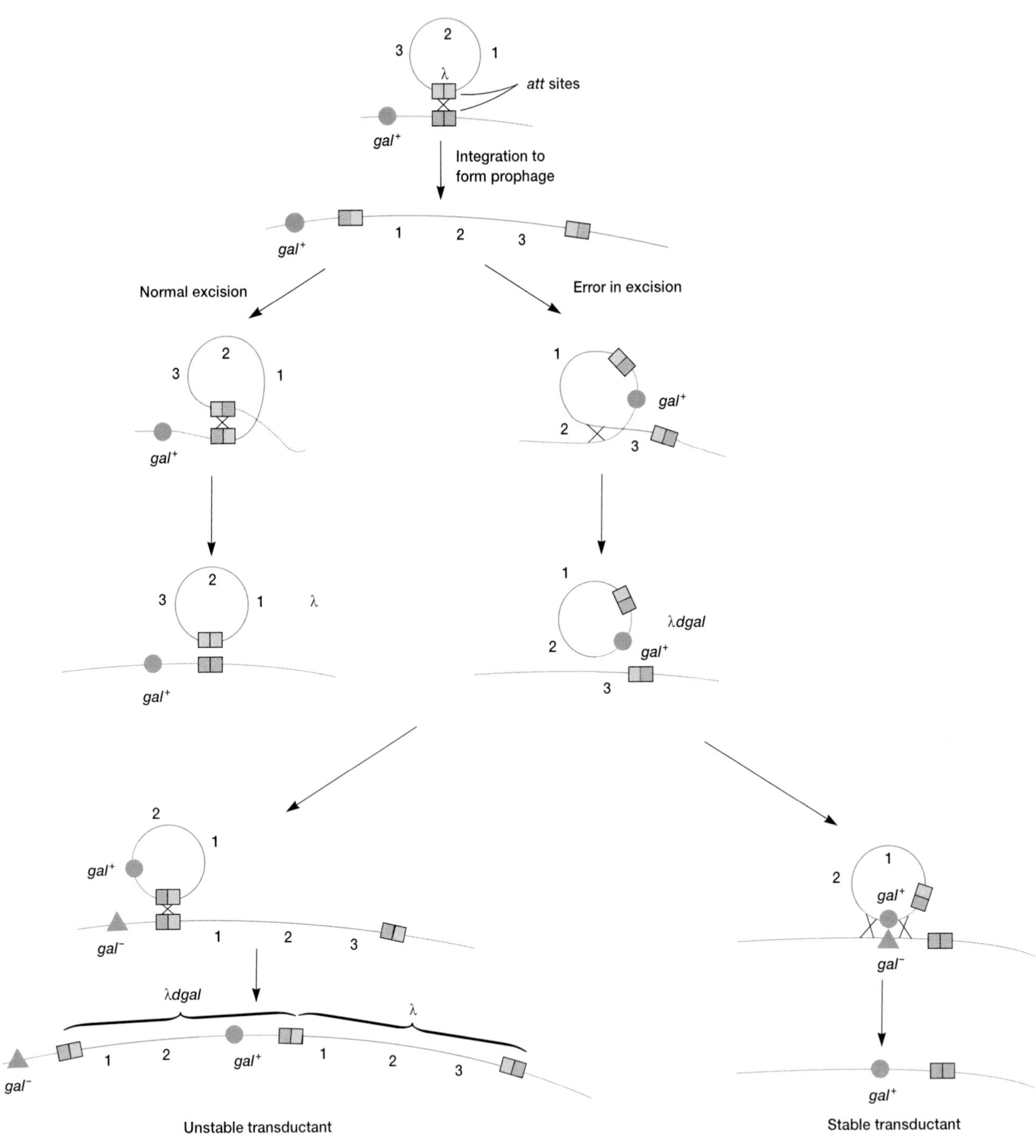

Mechanism of Transduction for Phage Lambda
Figure 14.20

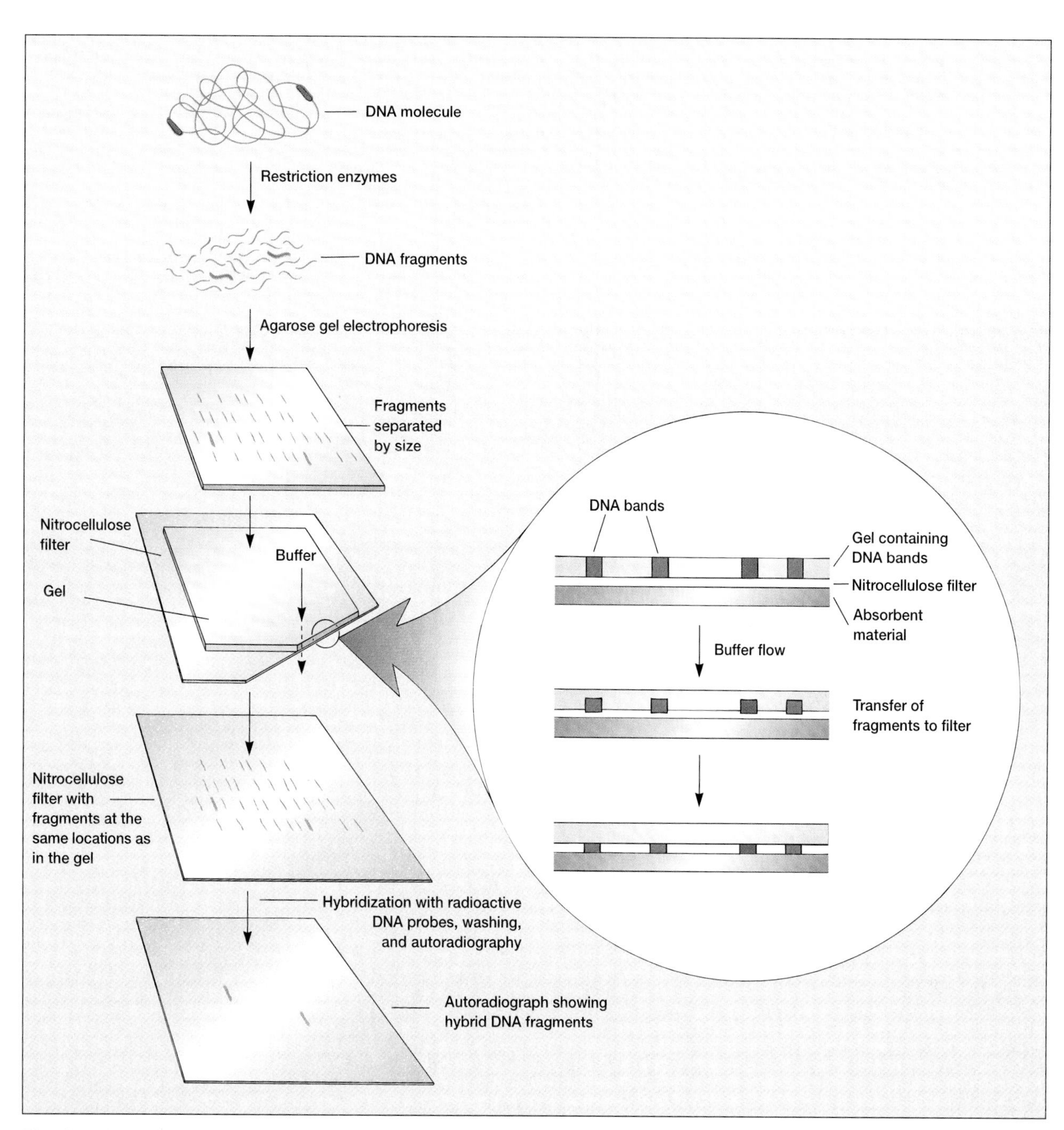
DNA molecule
Restriction enzymes
DNA fragments
Agarose gel electrophoresis
Fragments separated by size
Nitrocellulose filter
Buffer
Gel
Nitrocellulose filter with fragments at the same locations as in the gel
Hybridization with radioactive DNA probes, washing, and autoradiography
Autoradiograph showing hybrid DNA fragments
DNA bands
Gel containing DNA bands
Nitrocellulose filter
Absorbent material
Buffer flow
Transfer of fragments to filter

The Southern Blotting Technique
Figure 15.5

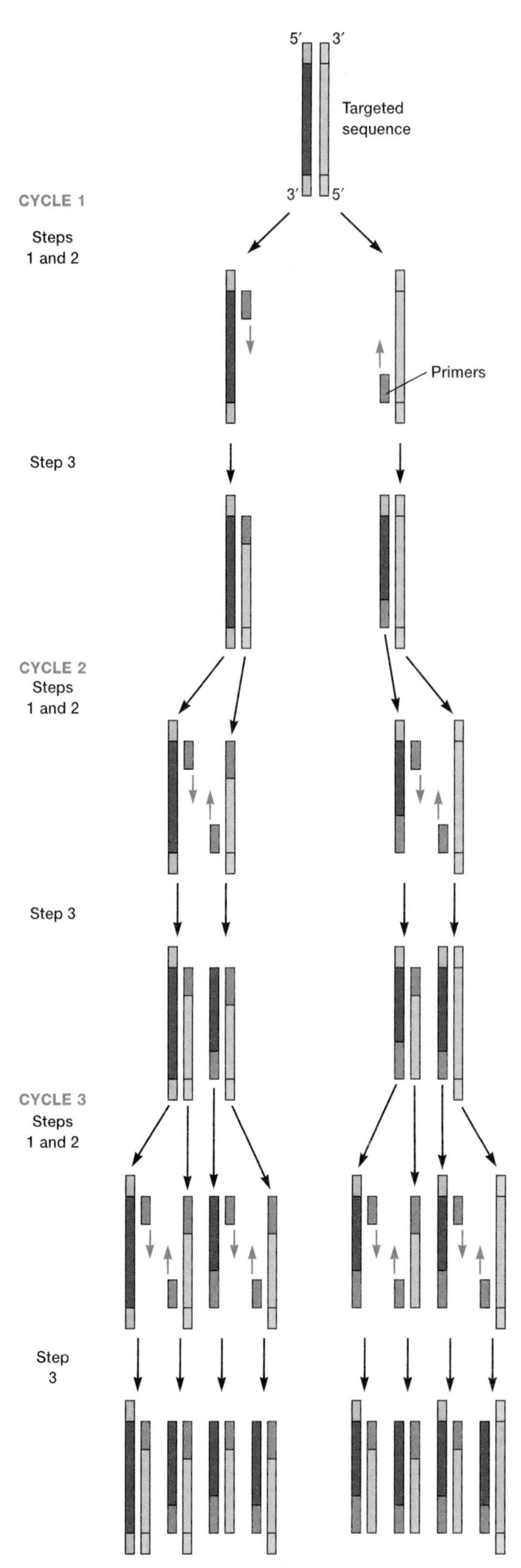
5′
3′
Targeted sequence
3′
5′
CYCLE 1
Steps 1 and 2
Primers
Step 3
CYCLE 2
Steps 1 and 2
Step 3
CYCLE 3
Steps 1 and 2
Step 3

The Polymerase Chain Reaction
Figure 15.8

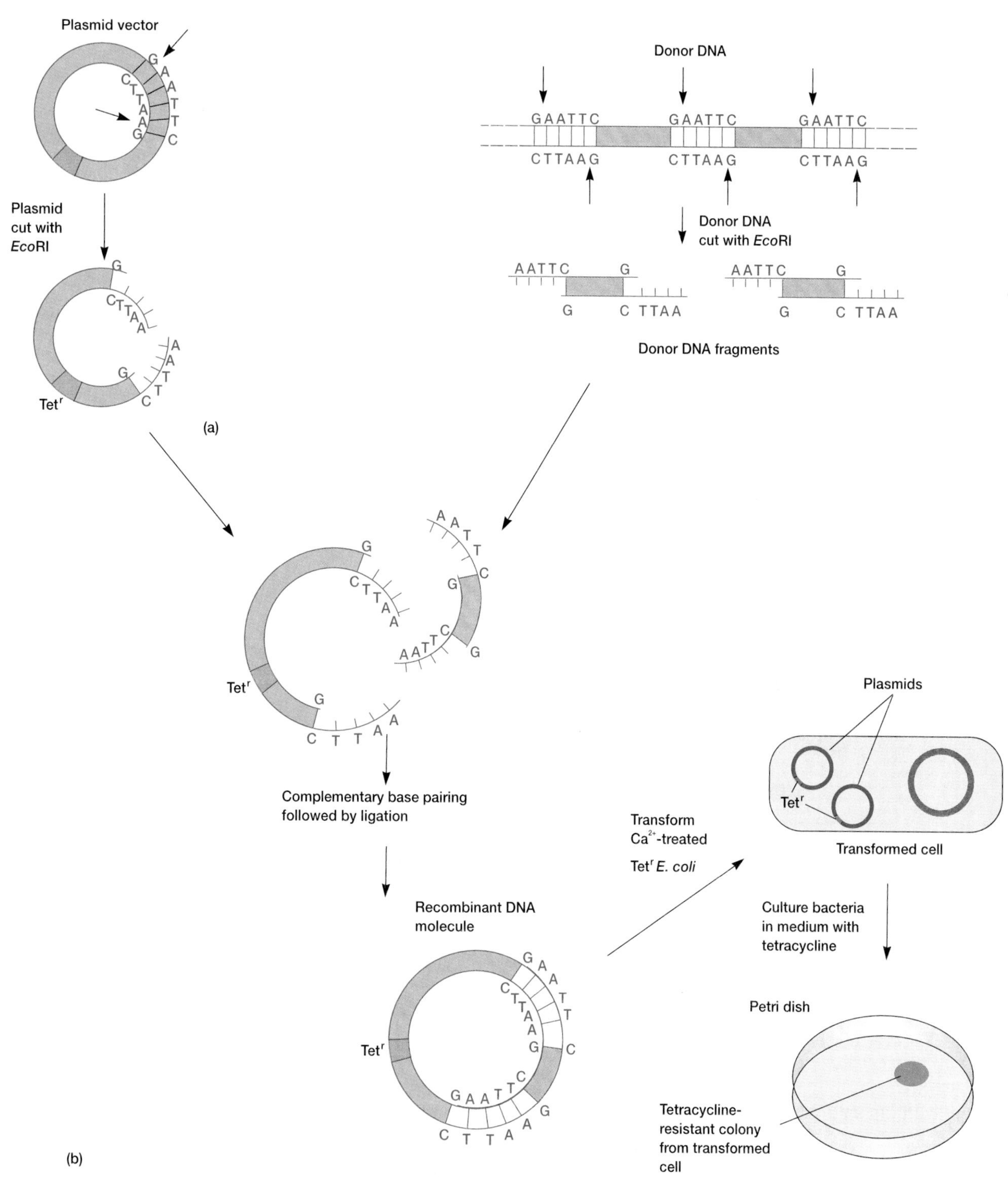

Recombinant Plasmid Construction and Cloning
Figure 15.11

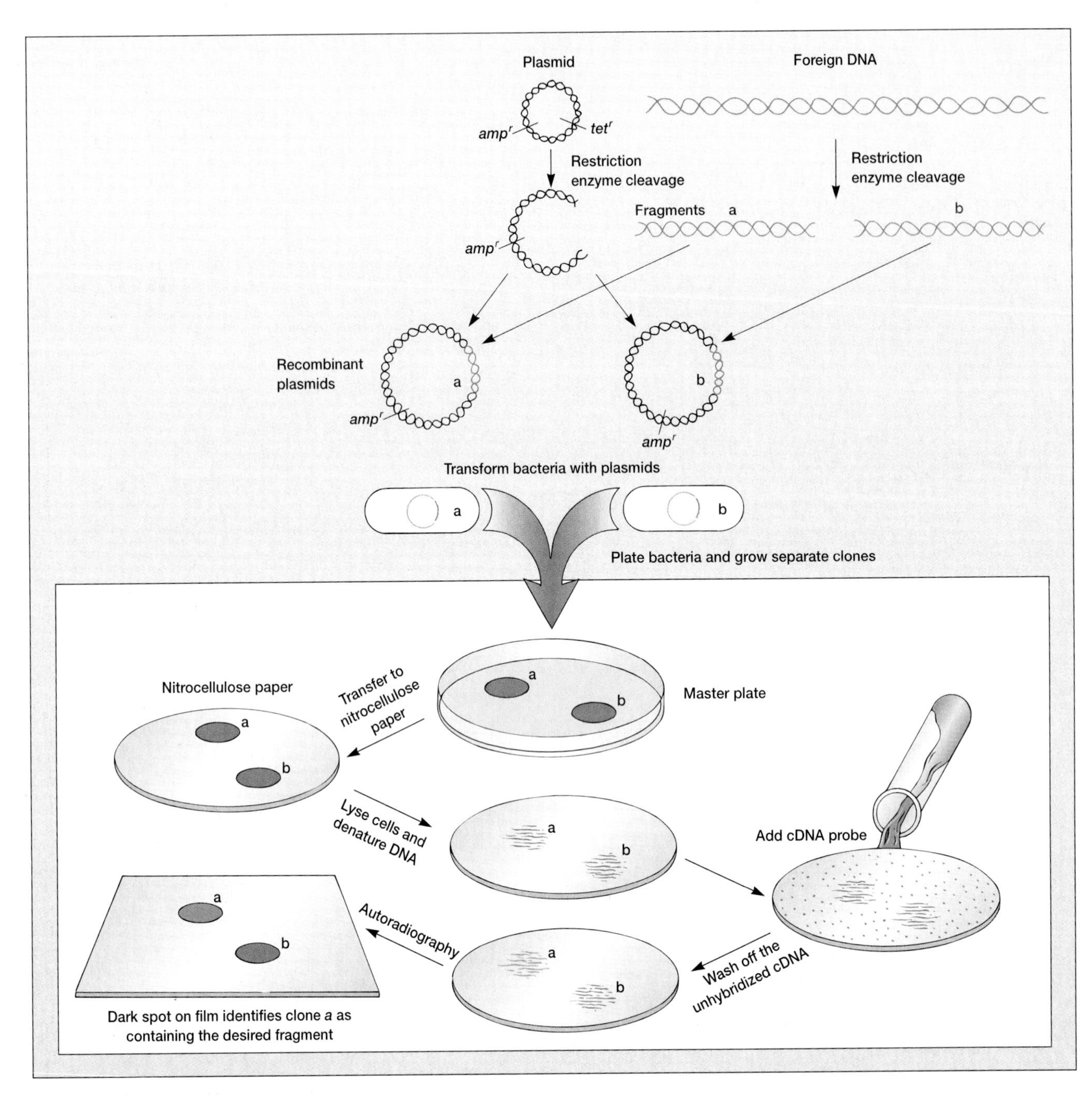

Plasmid

Foreign DNA

amp^r   tet^r

Restriction enzyme cleavage

Restriction enzyme cleavage

amp^r

Fragments   a   b

Recombinant plasmids

a   b

amp^r   amp^r

Transform bacteria with plasmids

a   b

Plate bacteria and grow separate clones

Nitrocellulose paper

a
b

Transfer to nitrocellulose paper

a
b

Master plate

Lyse cells and denature DNA

a
b

Add cDNA probe

a
b

Autoradiography

a
b

Wash off the unhybridized cDNA

a
b

Dark spot on film identifies clone *a* as containing the desired fragment

Cloning with Plasmid Vectors
Figure 15.14

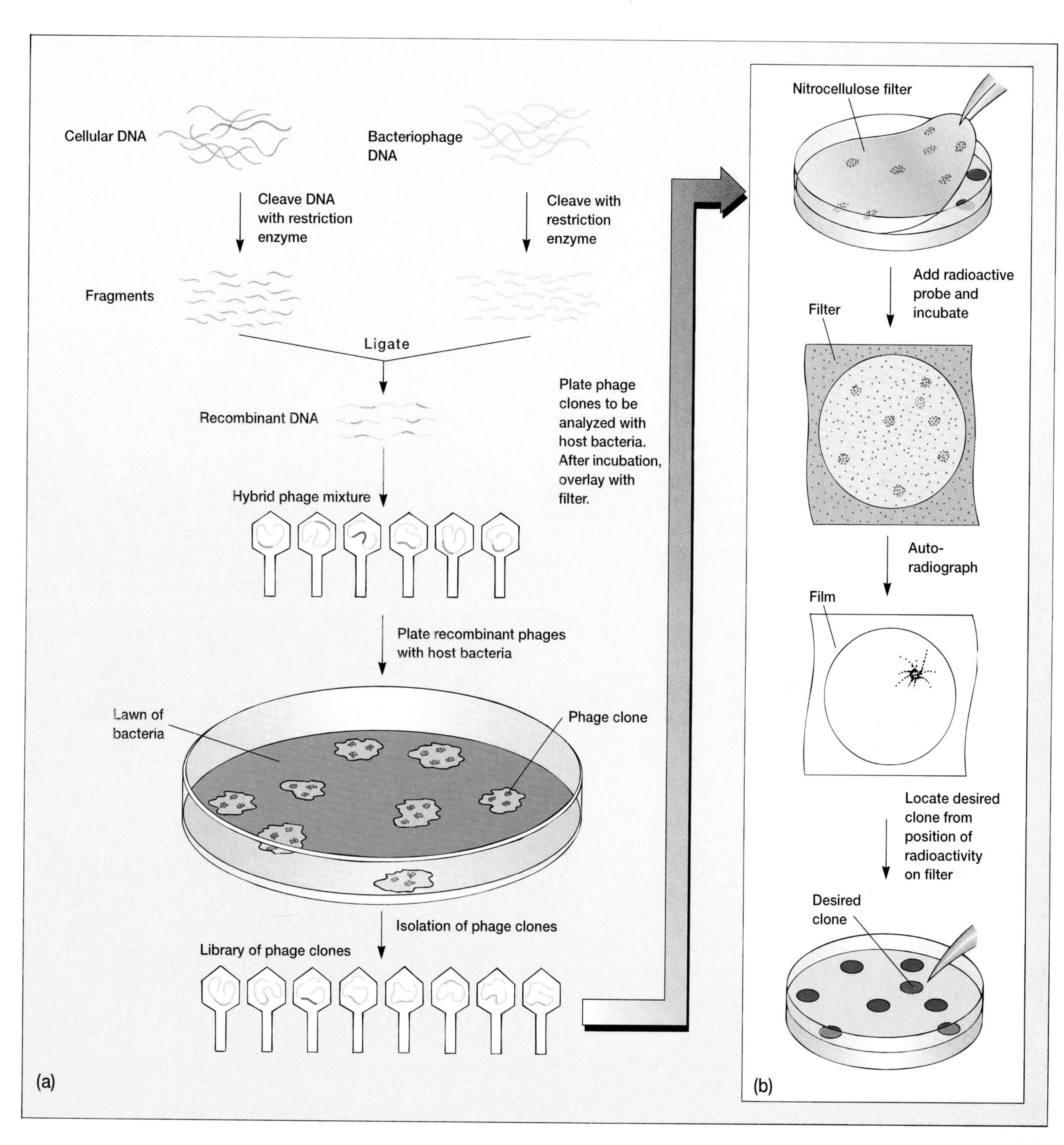
Cellular DNA
Bacteriophage DNA
Cleave DNA with restriction enzyme
Cleave with restriction enzyme
Fragments
Ligate
Recombinant DNA
Plate phage clones to be analyzed with host bacteria. After incubation, overlay with filter.
Hybrid phage mixture
Plate recombinant phages with host bacteria
Lawn of bacteria
Phage clone
Isolation of phage clones
Library of phage clones
(a)
Nitrocellulose filter
Add radioactive probe and incubate
Filter
Auto-radiograph
Film
Locate desired clone from position of radioactivity on filter
Desired clone
(b)

Use of Lambda Phage as a Vector
Figure 15.15

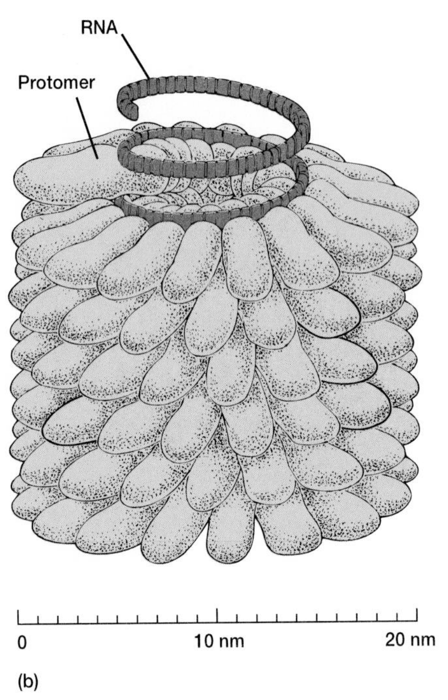

RNA

Protomer

0   10 nm   20 nm

(b)

Tobacco Mosaic Virus Structure
Figure 16.11 b

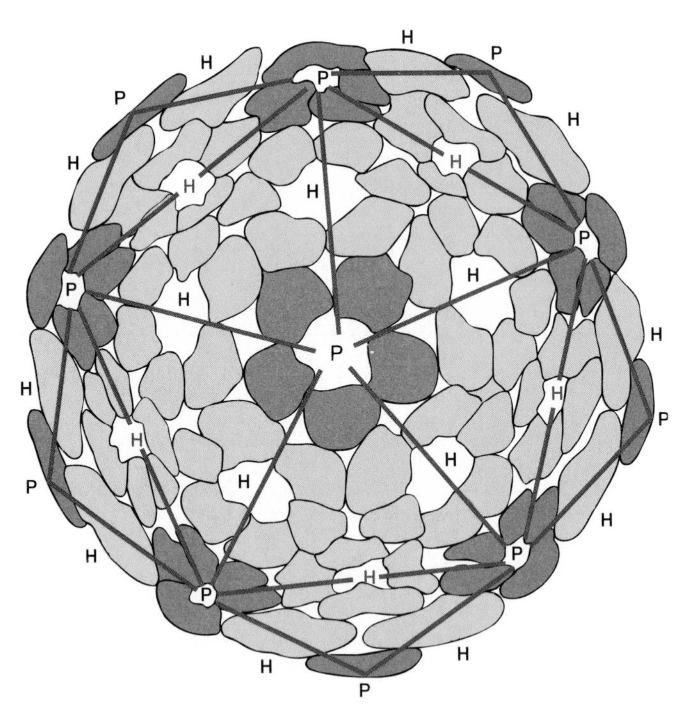

The Structure of an Icosahedral Capsid
Figure 16.13

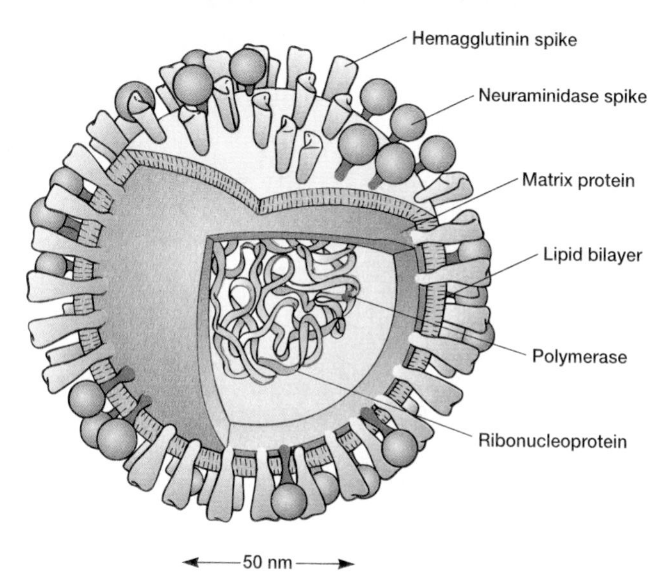

Hemagglutinin spike

Neuraminidase spike

Matrix protein

Lipid bilayer

Polymerase

Ribonucleoprotein

←—— 50 nm ——→

(b)

Influenza Virus
Figure 16.17 *b*

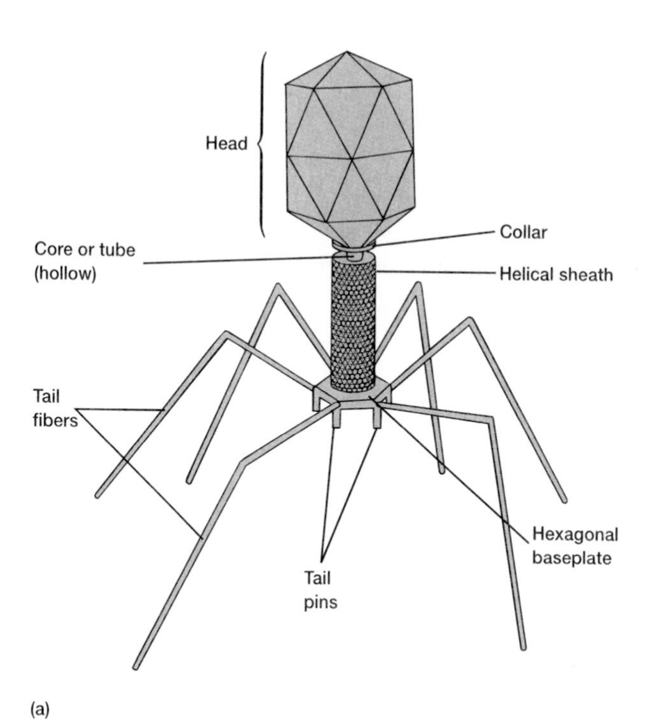

Head

Core or tube
(hollow)

Collar

Helical sheath

Tail
fibers

Tail
pins

Hexagonal
baseplate

(a)

T4 Coliphage Structure
Figure 16.19 *a*

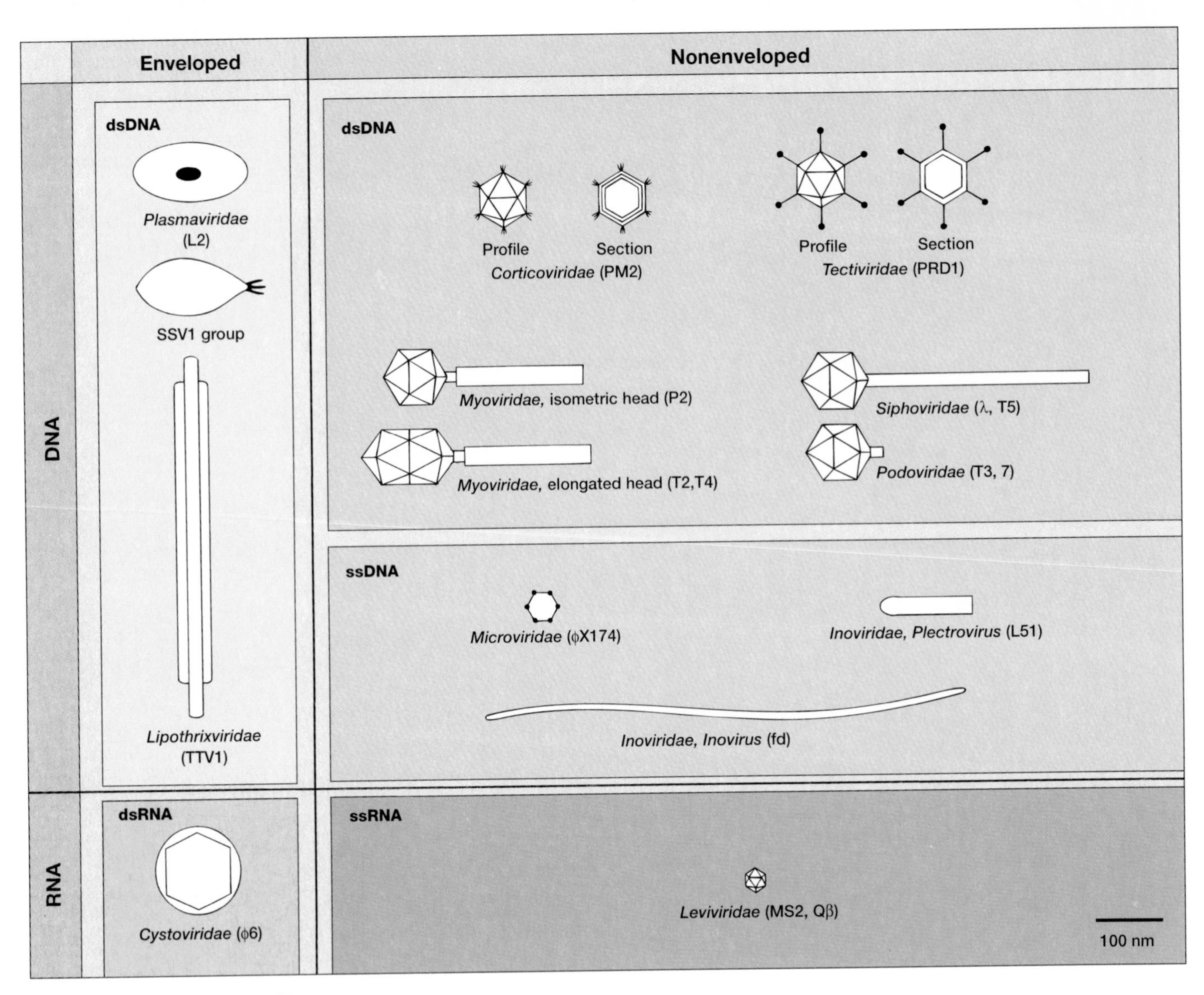

Major Bacteriophage Families
Figure 17.1

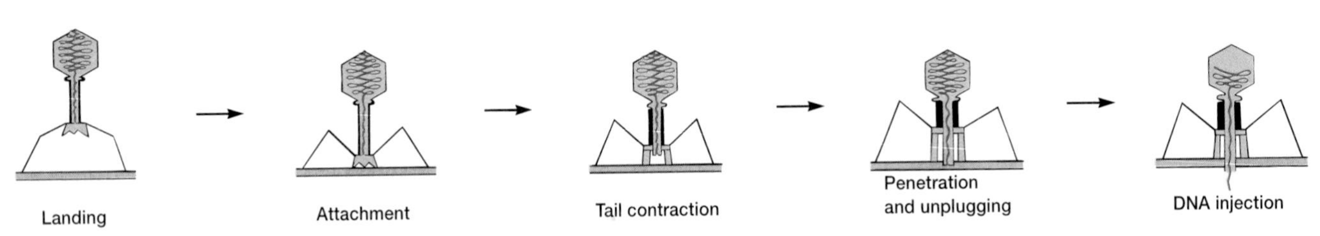

Landing → Attachment → Tail contraction → Penetration and unplugging → DNA injection

T4 Phage Adsorption and DNA Injection
Figure 17.3

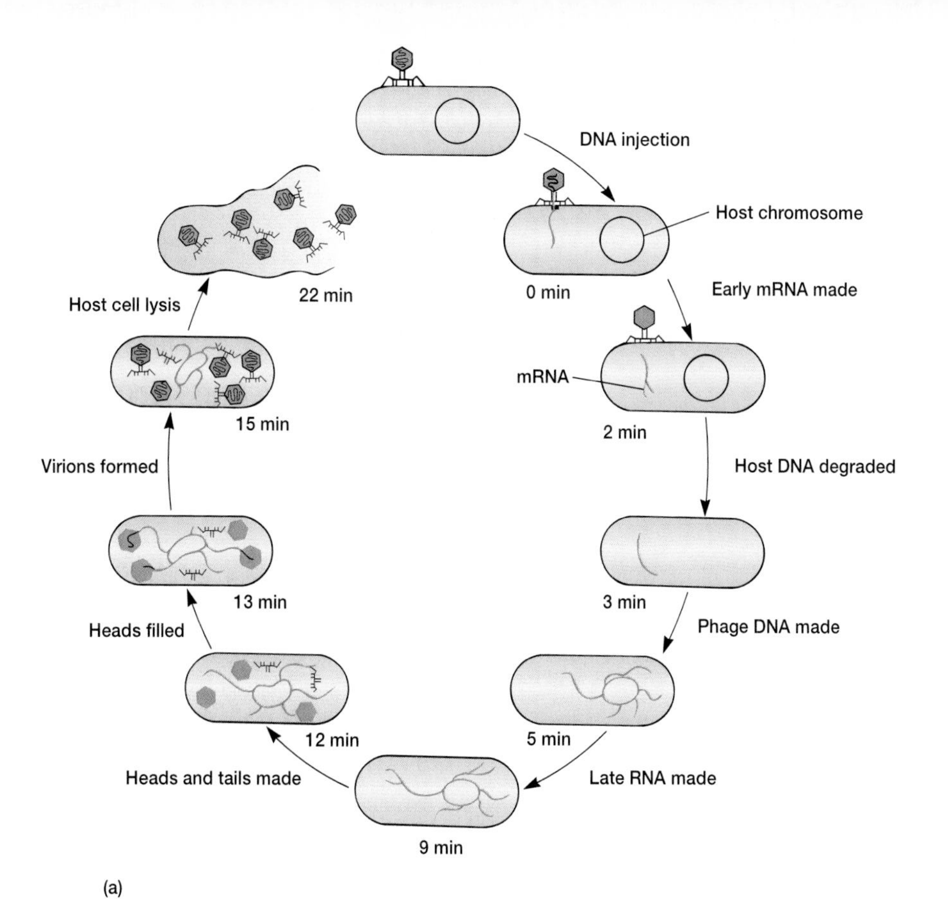

(a)

Life Cycle of Bacteriophage T4
Figure 17.5 *a*

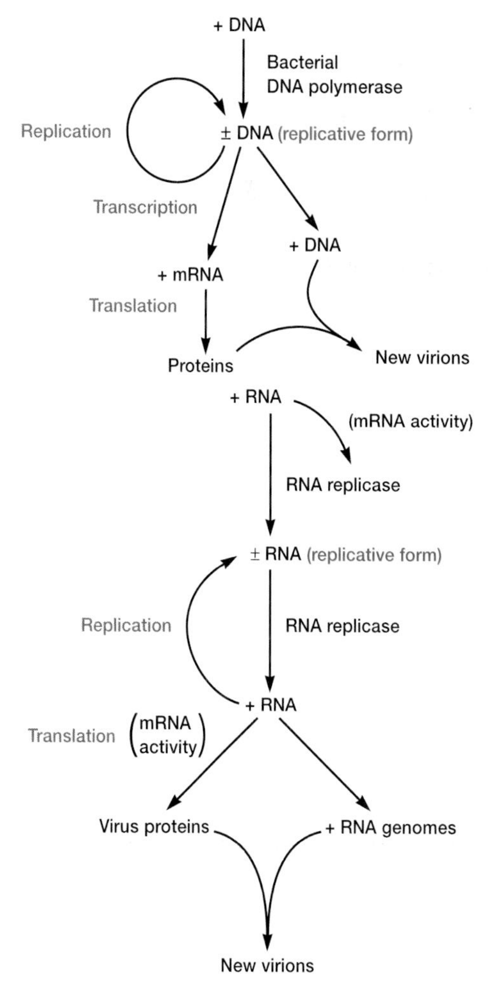

Reproduction of Single-Stranded Bacteriophages
Figure 17.10 & 17.13

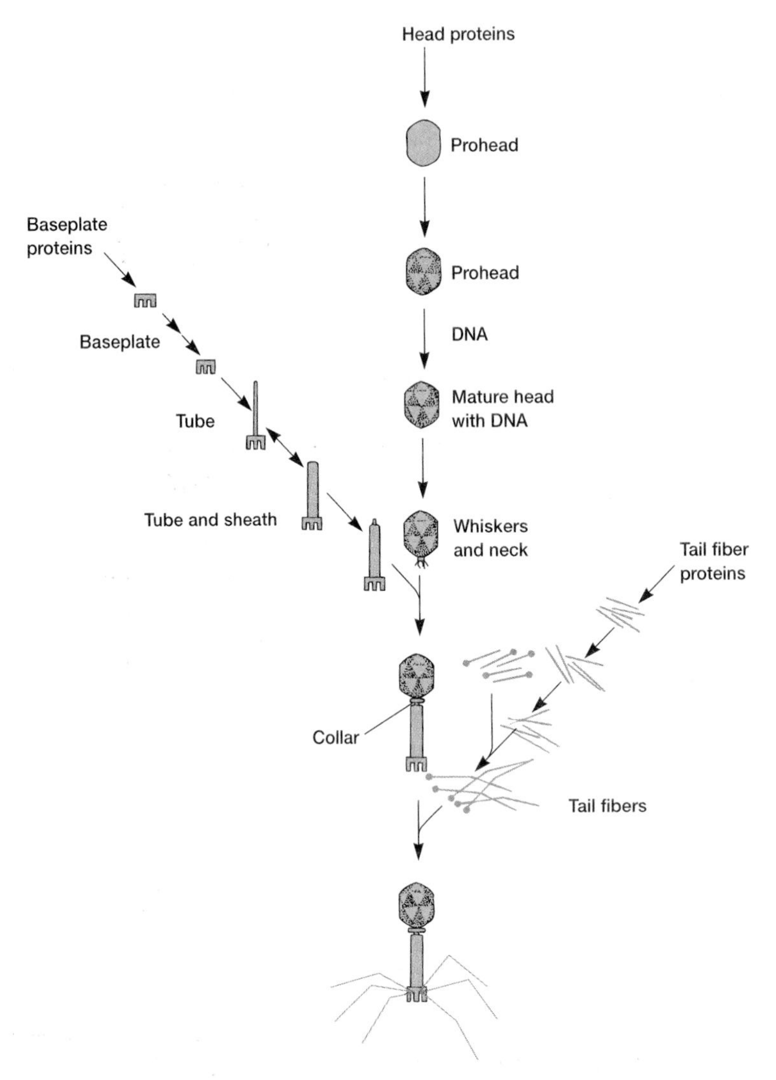

Head proteins

Prohead

Prohead

DNA

Mature head
with DNA

Whiskers
and neck

Baseplate
proteins

Baseplate

Tube

Tube and sheath

Collar

Tail fiber
proteins

Tail fibers

Assembly of T4 Bacteriophage
Figure 17.11

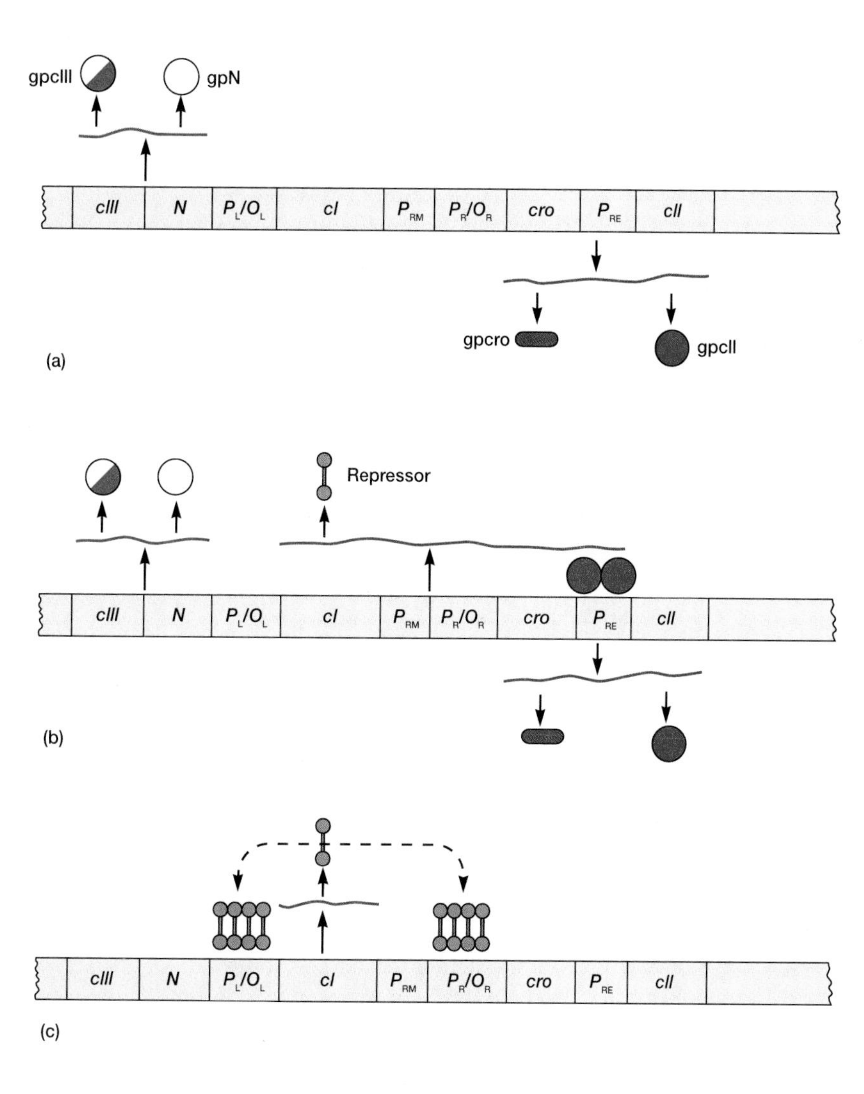
gpcIII
gpN
cIII
N
P_L/O_L
cI
P_RM
P_R/O_R
cro
P_RE
cII
gpcro
gpcII
(a)
Repressor
(b)
(c)

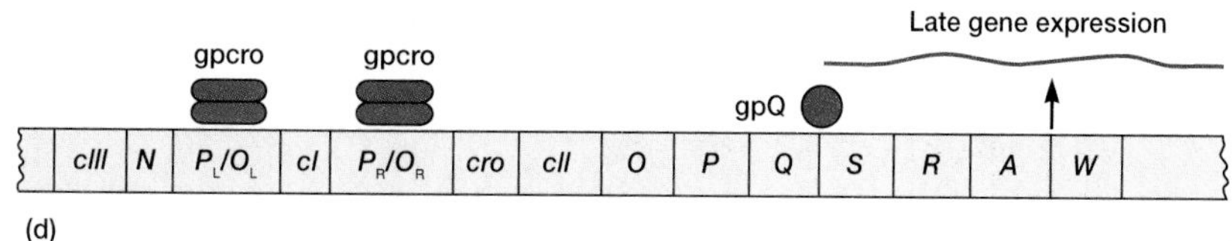
Late gene expression
gpcro
gpcro
gpQ
cIII
N
P_L/O_L
cI
P_R/O_R
cro
cII
O
P
Q
S
R
A
W
(d)

Choice Between Lysogeny and Lysis
Figure 17.18

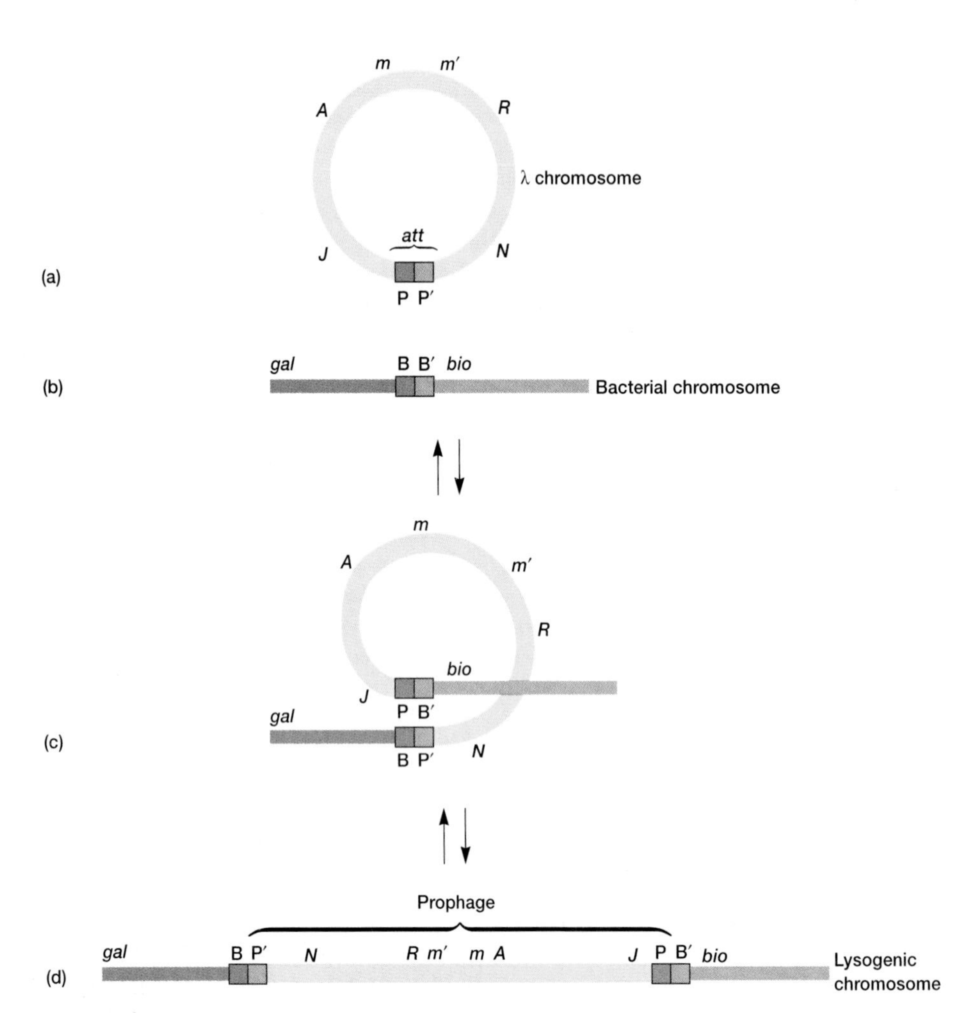

(a)

(b)

(c)

(d)

Insertion and Excision of Lambda Phage
Figure 17.20

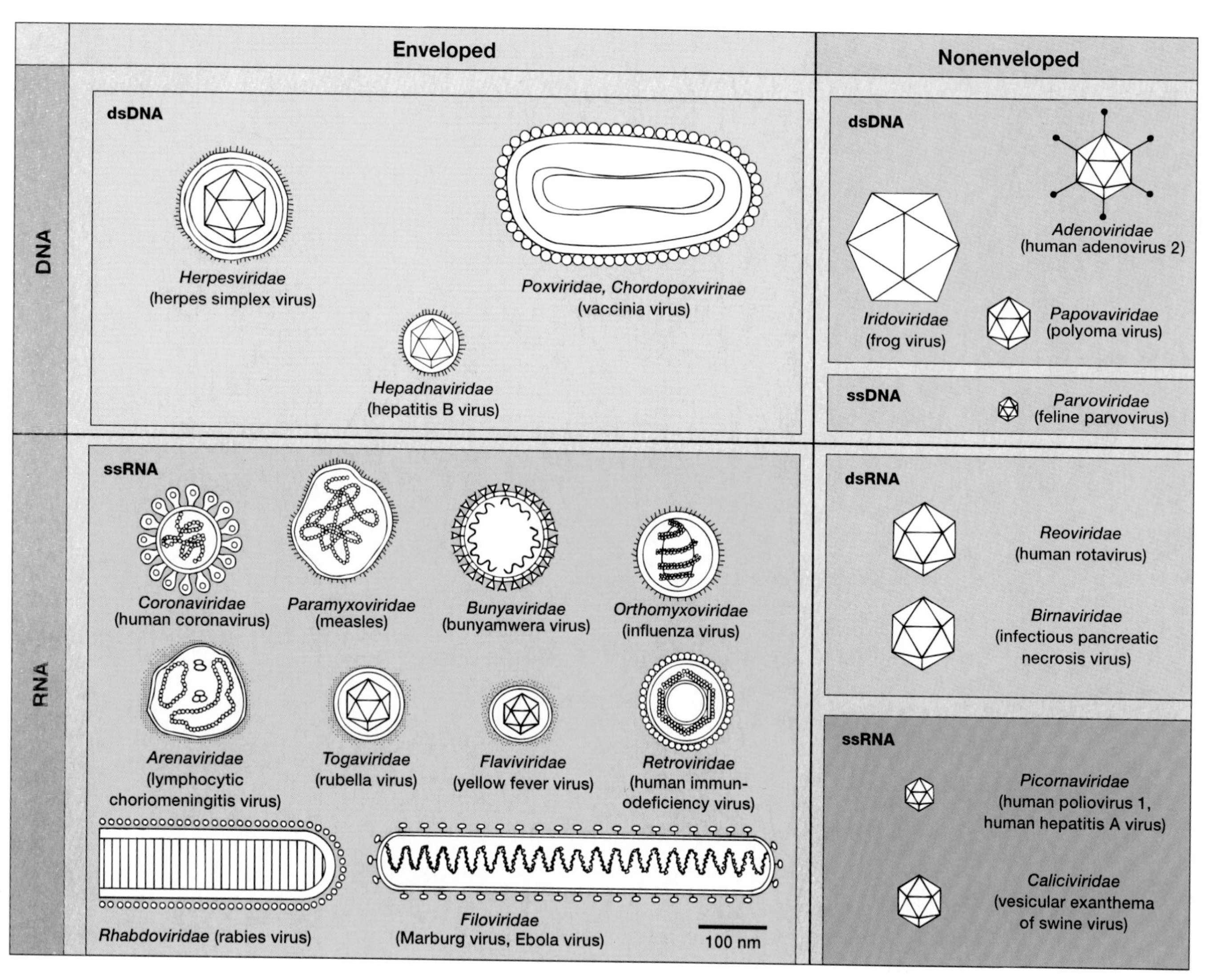

Enveloped
Nonenveloped
DNA
RNA
dsDNA
Herpesviridae
(herpes simplex virus)
Poxviridae, Chordopoxvirinae
(vaccinia virus)
Hepadnaviridae
(hepatitis B virus)
dsDNA
Adenoviridae
(human adenovirus 2)
Iridoviridae
(frog virus)
Papovaviridae
(polyoma virus)
ssDNA
Parvoviridae
(feline parvovirus)
ssRNA
Coronaviridae
(human coronavirus)
Paramyxoviridae
(measles)
Bunyaviridae
(bunyamwera virus)
Orthomyxoviridae
(influenza virus)
Arenaviridae
(lymphocytic
choriomeningitis virus)
Togaviridae
(rubella virus)
Flaviviridae
(yellow fever virus)
Retroviridae
(human immun-
odeficiency virus)
Rhabdoviridae (rabies virus)
Filoviridae
(Marburg virus, Ebola virus)
100 nm
dsRNA
Reoviridae
(human rotavirus)
Birnaviridae
(infectious pancreatic
necrosis virus)
ssRNA
Picornaviridae
(human poliovirus 1,
human hepatitis A virus)
Caliciviridae
(vesicular exanthema
of swine virus)

The Major Animal Virus Families
Figure 18.3

(a) Direct penetration by naked viruses

Capsid

Nucleic acid

Receptor

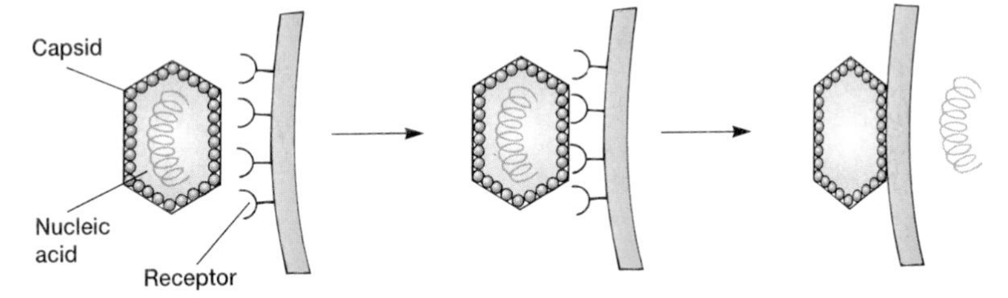

(b) Enveloped virus fusing with plasma membrane

Capsid protein

Spikes

Nucleic acid

Envelope

Nucleic acid

Capsid

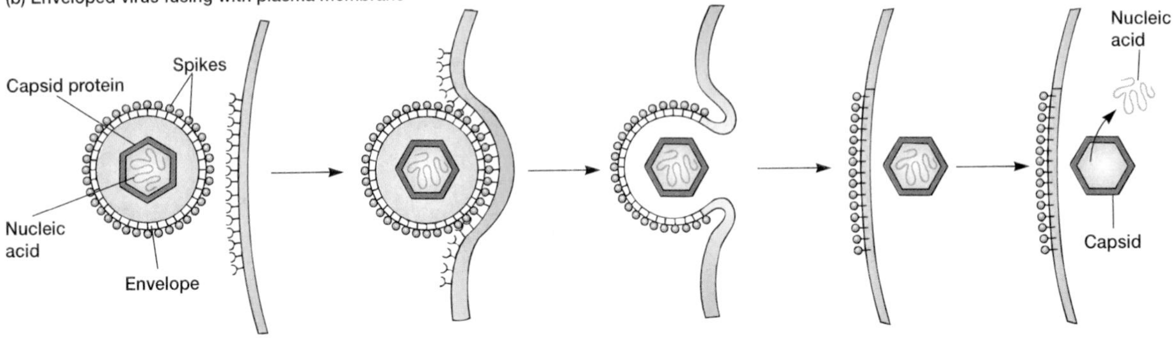

(c) Entry of enveloped virus by endocytosis

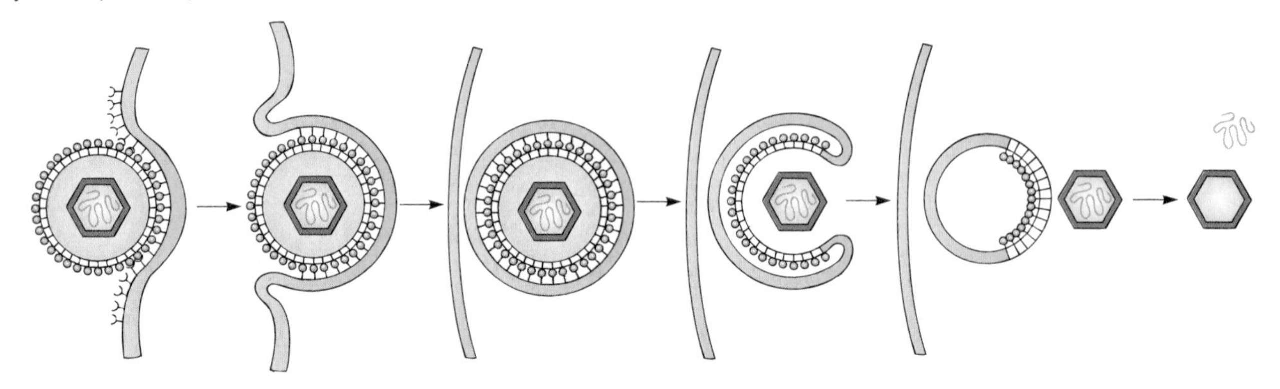

Animal Virus Entry
Figure 18.4

(a) Picornaviruses (poliovirus)

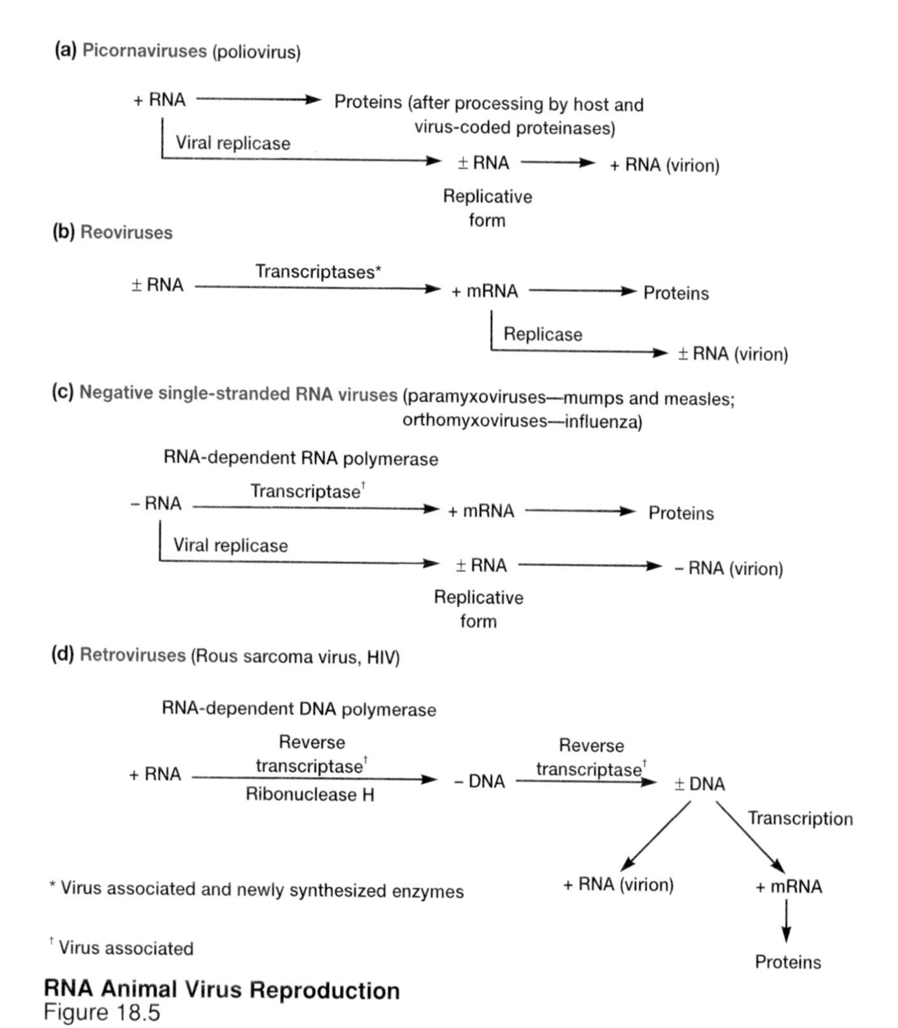

(b) Reoviruses

(c) Negative single-stranded RNA viruses (paramyxoviruses—mumps and measles; orthomyxoviruses—influenza)

(d) Retroviruses (Rous sarcoma virus, HIV)

* Virus associated and newly synthesized enzymes

† Virus associated

RNA Animal Virus Reproduction
Figure 18.5

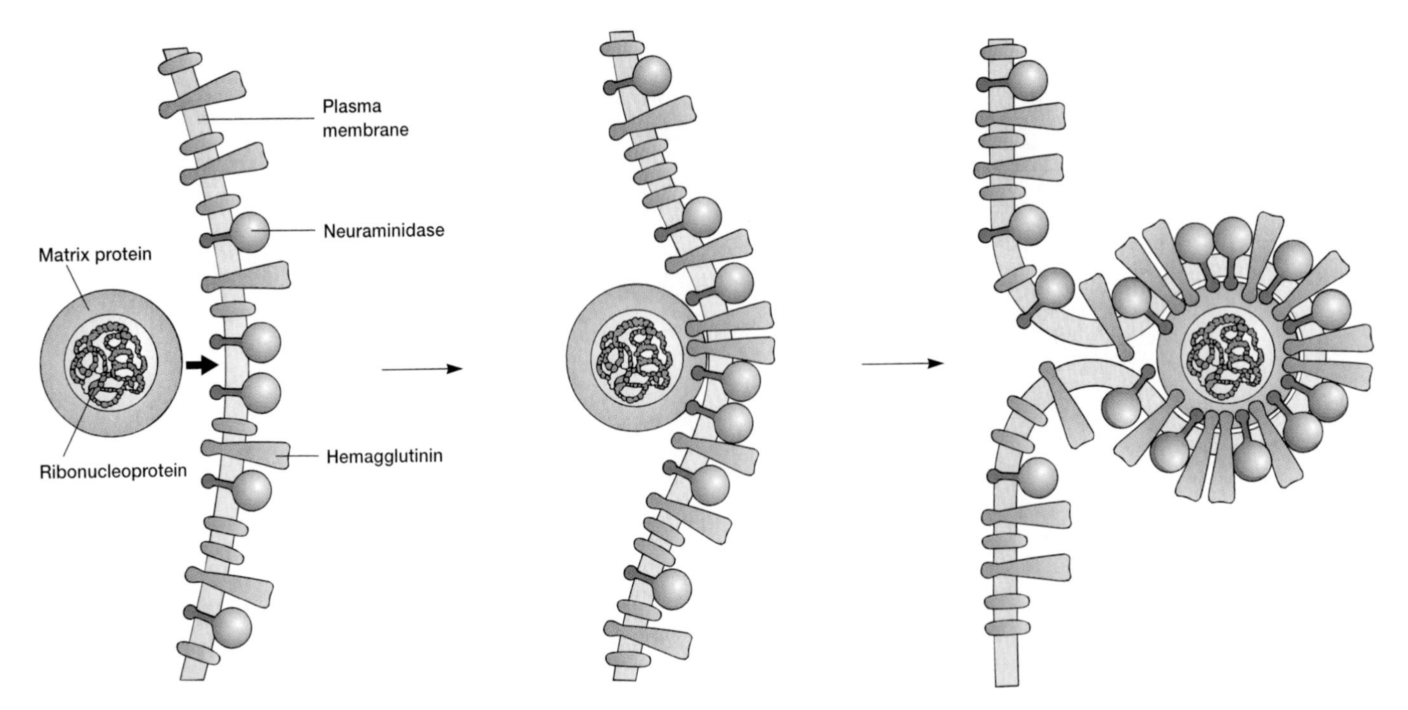

Release of Influenza Virus
Figure 18.8

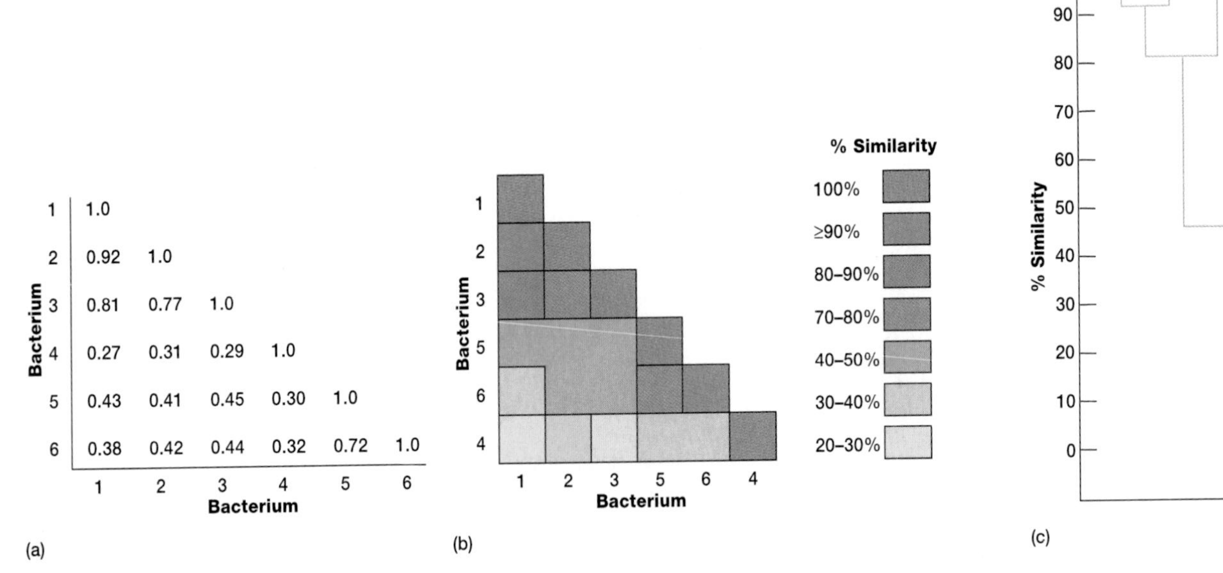

(a)

(b)

% Similarity

100%	
≥90%	
80–90%	
70–80%	
40–50%	
30–40%	
20–30%	

(c)

Clustering and Dendrograms
Figure 19.4

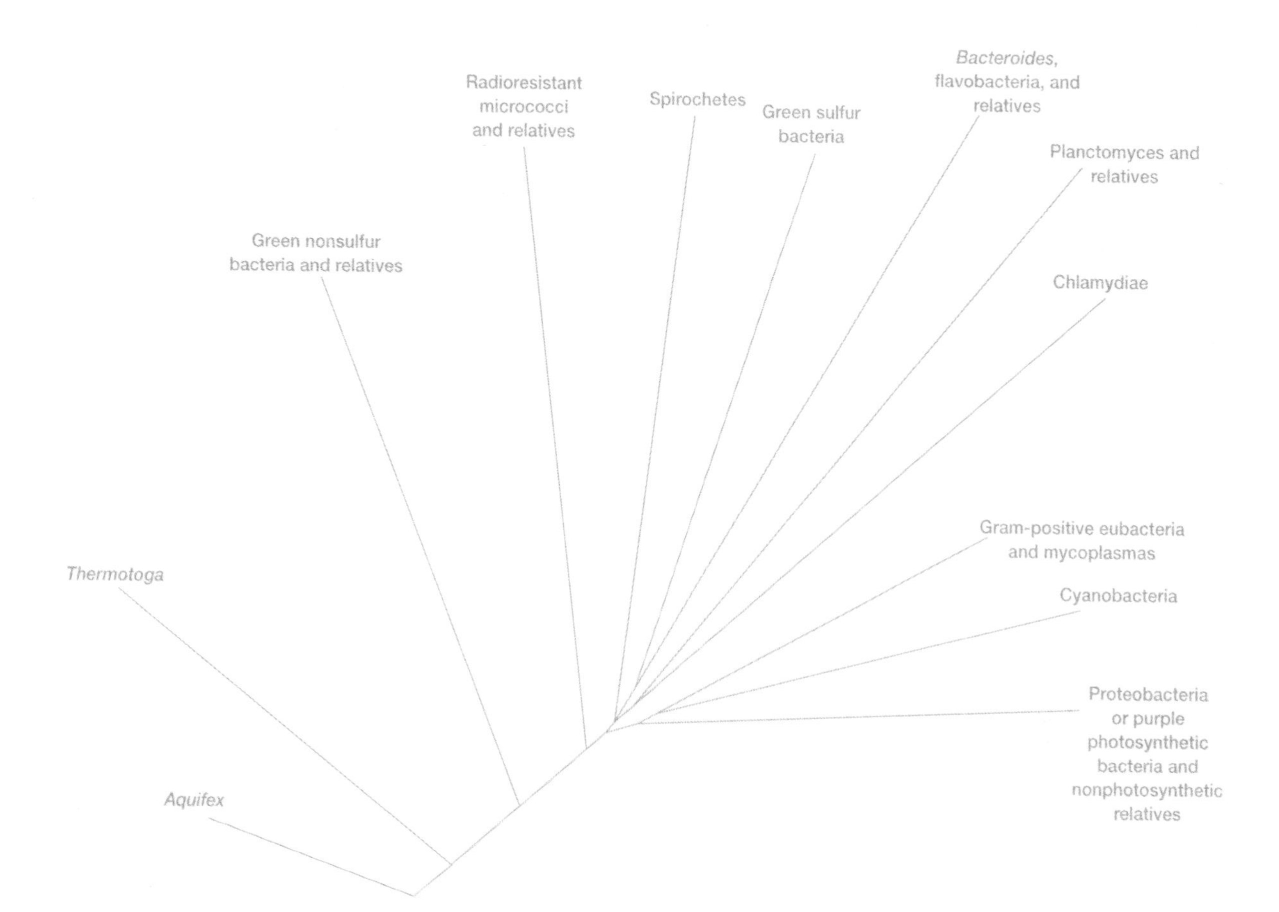
Radioresistant micrococci and relatives
Spirochetes
Green sulfur bacteria
Bacteroides, flavobacteria, and relatives
Planctomyces and relatives
Green nonsulfur bacteria and relatives
Chlamydiae
Gram-positive eubacteria and mycoplasmas
Thermotoga
Cyanobacteria
Proteobacteria or purple photosynthetic bacteria and nonphotosynthetic relatives
Aquifex

Eubacterial Phylogenetic Tree
Figure 19.7

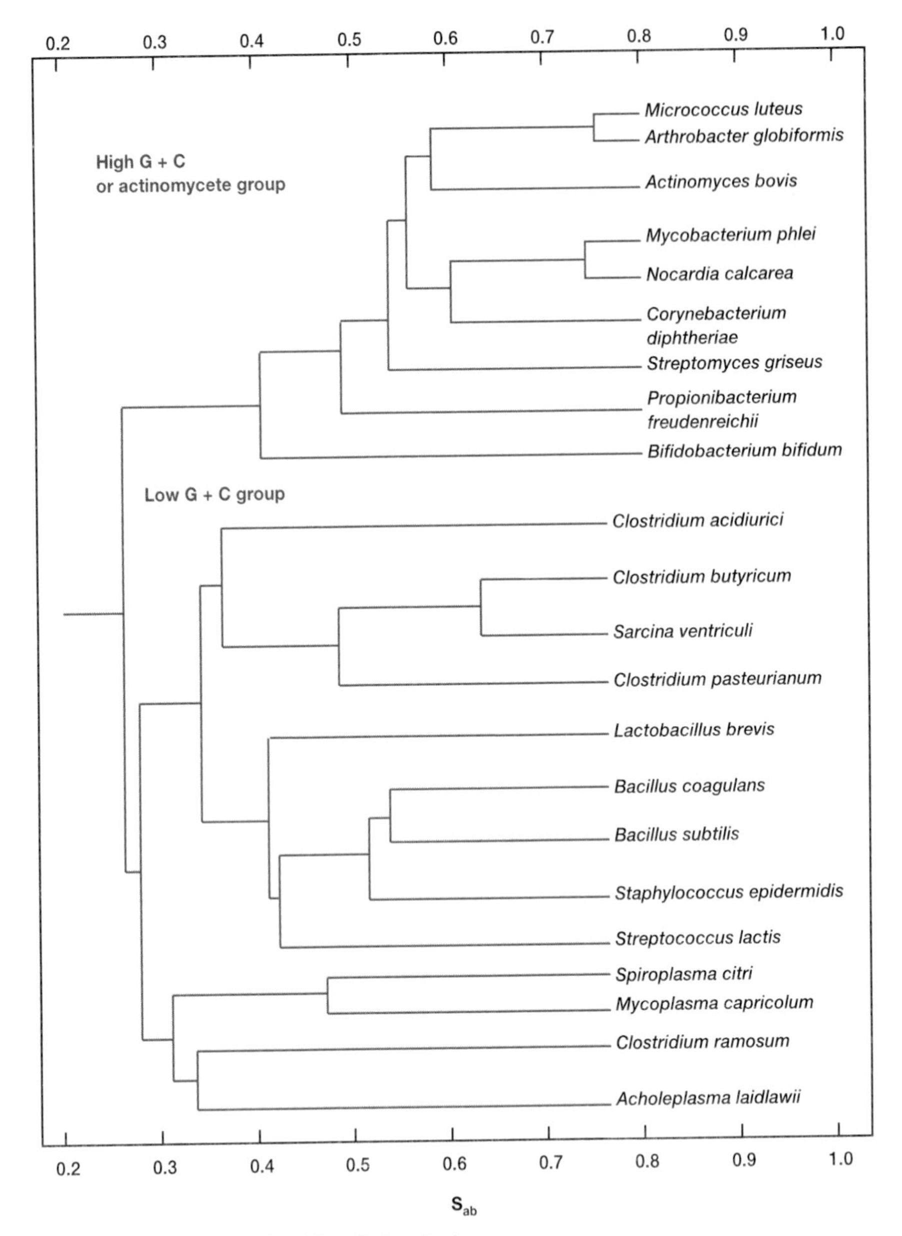

Phylogeny of Gram-Positive Eubacteria
Figure 19.8

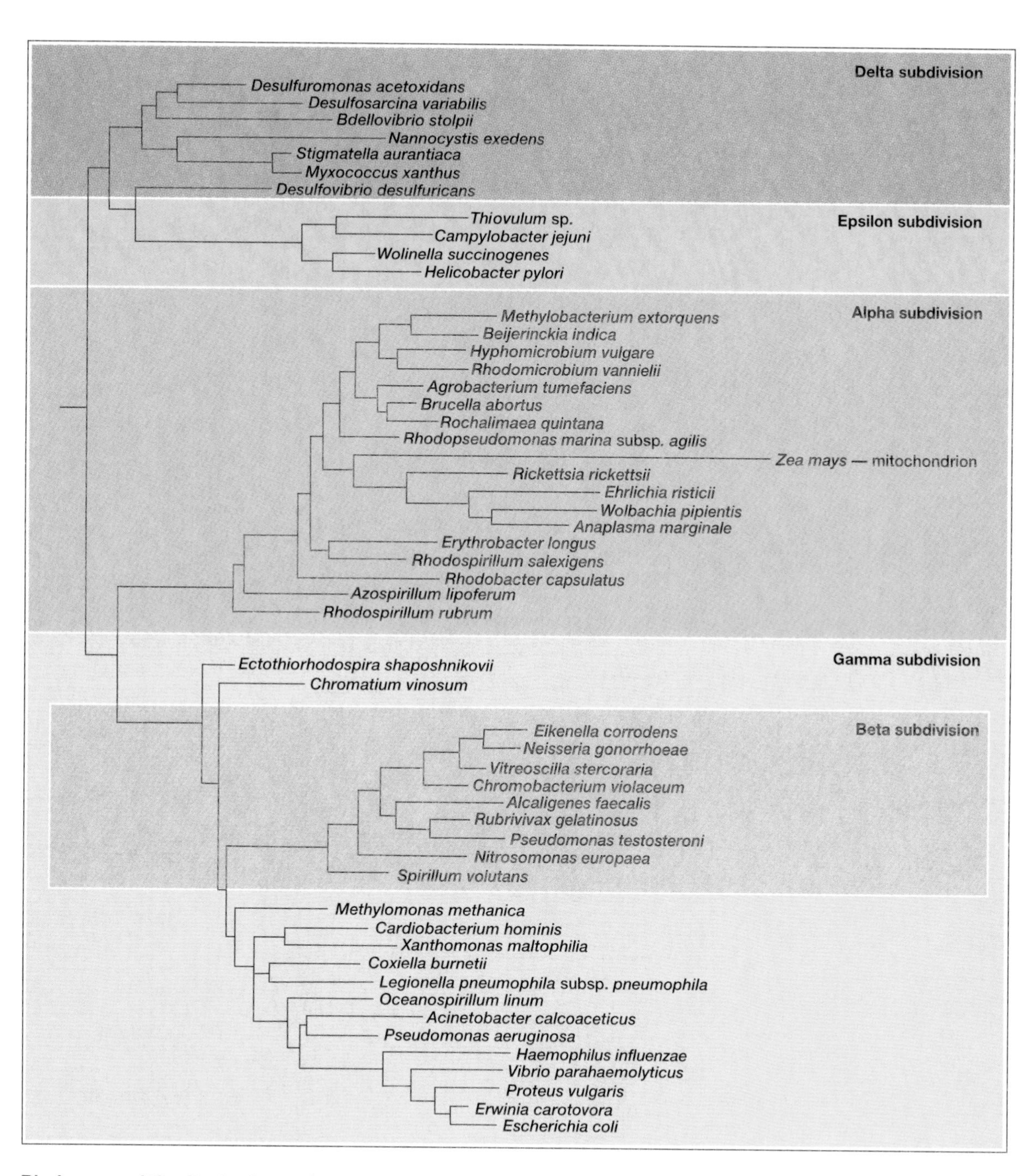

Delta subdivision
Desulfuromonas acetoxidans
Desulfosarcina variabilis
Bdellovibrio stolpii
Nannocystis exedens
Stigmatella aurantiaca
Myxococcus xanthus
Desulfovibrio desulfuricans
Epsilon subdivision
Thiovulum sp.
Campylobacter jejuni
Wolinella succinogenes
Helicobacter pylori
Alpha subdivision
Methylobacterium extorquens
Beijerinckia indica
Hyphomicrobium vulgare
Rhodomicrobium vannielii
Agrobacterium tumefaciens
Brucella abortus
Rochalimaea quintana
Rhodopseudomonas marina subsp. agilis
Zea mays — mitochondrion
Rickettsia rickettsii
Ehrlichia risticii
Wolbachia pipientis
Anaplasma marginale
Erythrobacter longus
Rhodospirillum salexigens
Rhodobacter capsulatus
Azospirillum lipoferum
Rhodospirillum rubrum
Gamma subdivision
Ectothiorhodospira shaposhnikovii
Chromatium vinosum
Beta subdivision
Eikenella corrodens
Neisseria gonorrhoeae
Vitreoscilla stercoraria
Chromobacterium violaceum
Alcaligenes faecalis
Rubrivivax gelatinosus
Pseudomonas testosteroni
Nitrosomonas europaea
Spirillum volutans
Methylomonas methanica
Cardiobacterium hominis
Xanthomonas maltophilia
Coxiella burnetii
Legionella pneumophila subsp. pneumophila
Oceanospirillum linum
Acinetobacter calcoaceticus
Pseudomonas aeruginosa
Haemophilus influenzae
Vibrio parahaemolyticus
Proteus vulgaris
Erwinia carotovora
Escherichia coli

Phylogeny of the Proteobacteria
Figure 19.9

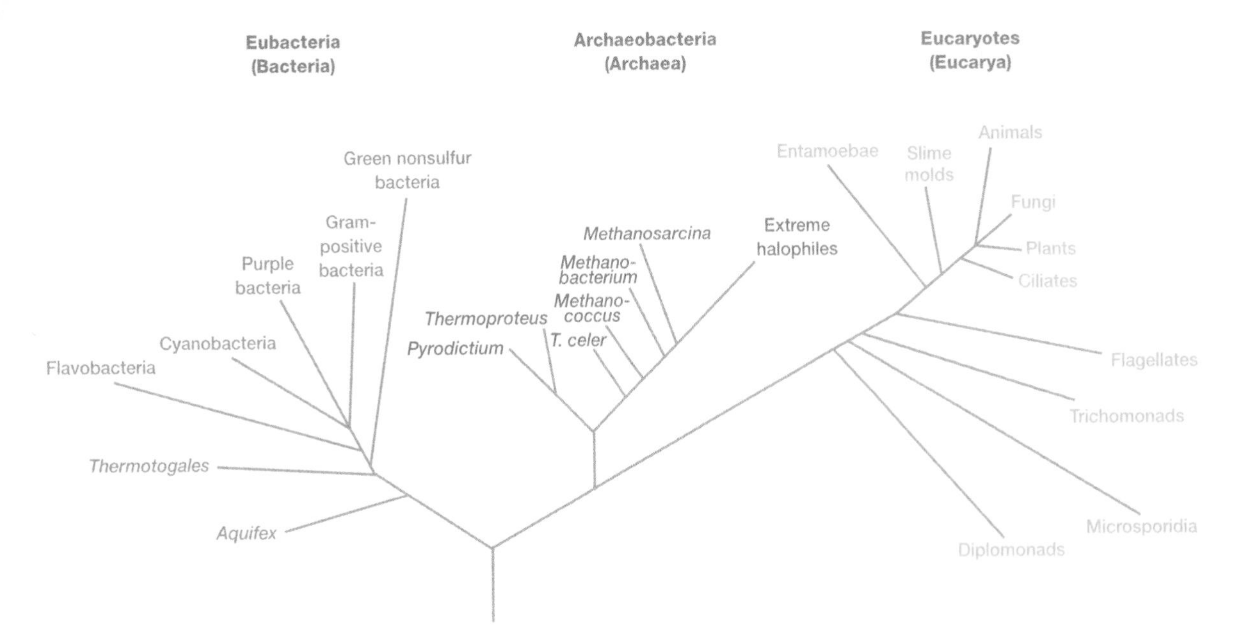

Eubacteria
(Bacteria)

Archaeobacteria
(Archaea)

Eucaryotes
(Eucarya)

Green nonsulfur
bacteria

Gram-
positive
bacteria

Purple
bacteria

Methanosarcina

Extreme
halophiles

Entamoebae

Slime
molds

Animals

Fungi

Cyanobacteria

*Methano-
bacterium*

*Methano-
coccus*

Plants

Thermoproteus

Pyrodictium

T. celer

Ciliates

Flavobacteria

Flagellates

Thermotogales

Trichomonads

Aquifex

Microsporidia

Diplomonads

Universal Phylogenetic Tree
Figure 19.13

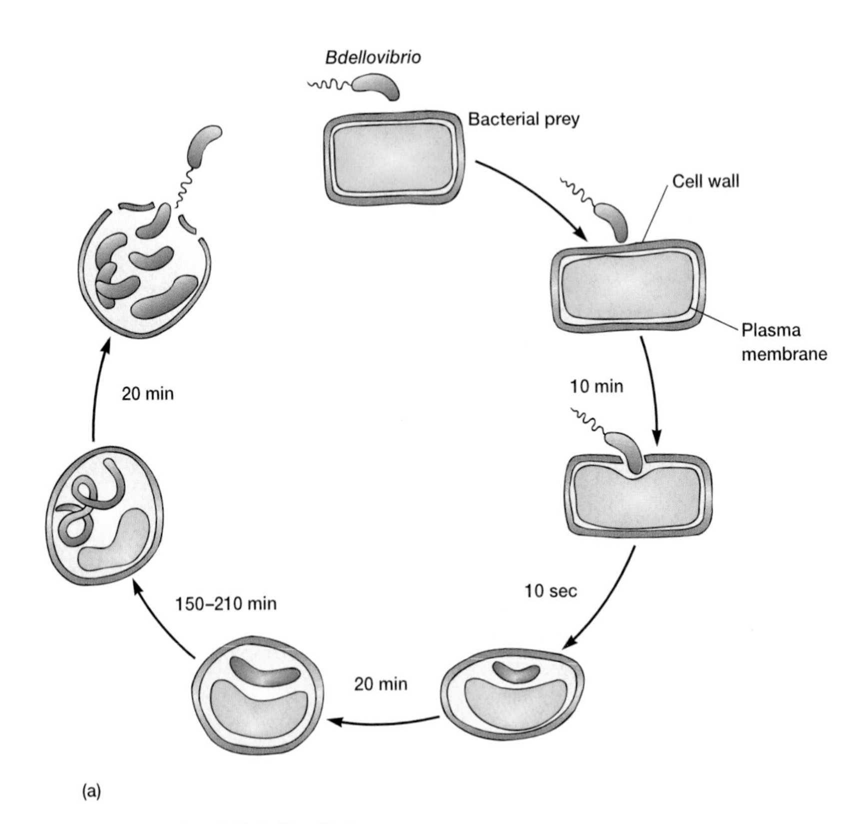

Bdellovibrio

Bacterial prey

Cell wall

Plasma
membrane

20 min

10 min

150–210 min

10 sec

20 min

(a)

The Life Cycle of *Bdellovibrio*
Figure 20.7 *a*

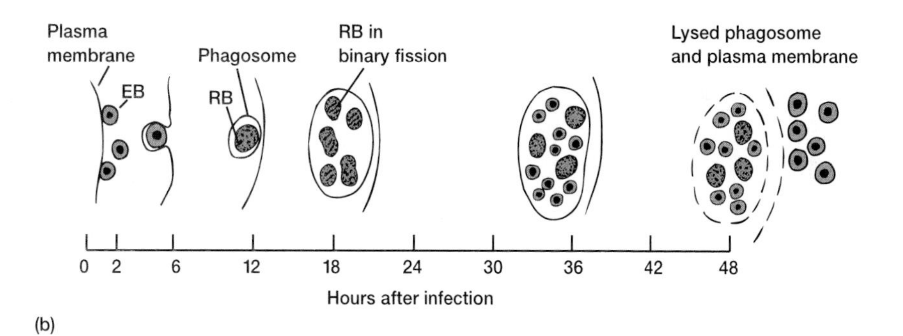

Elementary body

Size about 0.3 μm
Rigid cell wall
Relatively resistant to sonication
Resistant to trypsin
Subunit in cell envelope
RNA:DNA content = 1:1
Toxic for mice
Isolated organisms infectious
Adapted for extracellular survival

Reticulate body (initial body)

Size 0.5–1.0 μm
Fragile cell wall
Sensitive to sonication
Lysed by trypsin
No subunit in envelope
RNA:DNA content = 3:1
Nontoxic for mice
Isolated organisms not infectious
Adapted for intracellular growth

Plasma membrane

Phagosome

RB in binary fission

Lysed phagosome and plasma membrane

EB

RB

0 2   6   12   18   24   30   36   42   48

Hours after infection

(b)

The Chlamydial Life Cycle
Figure 20.20 *b*

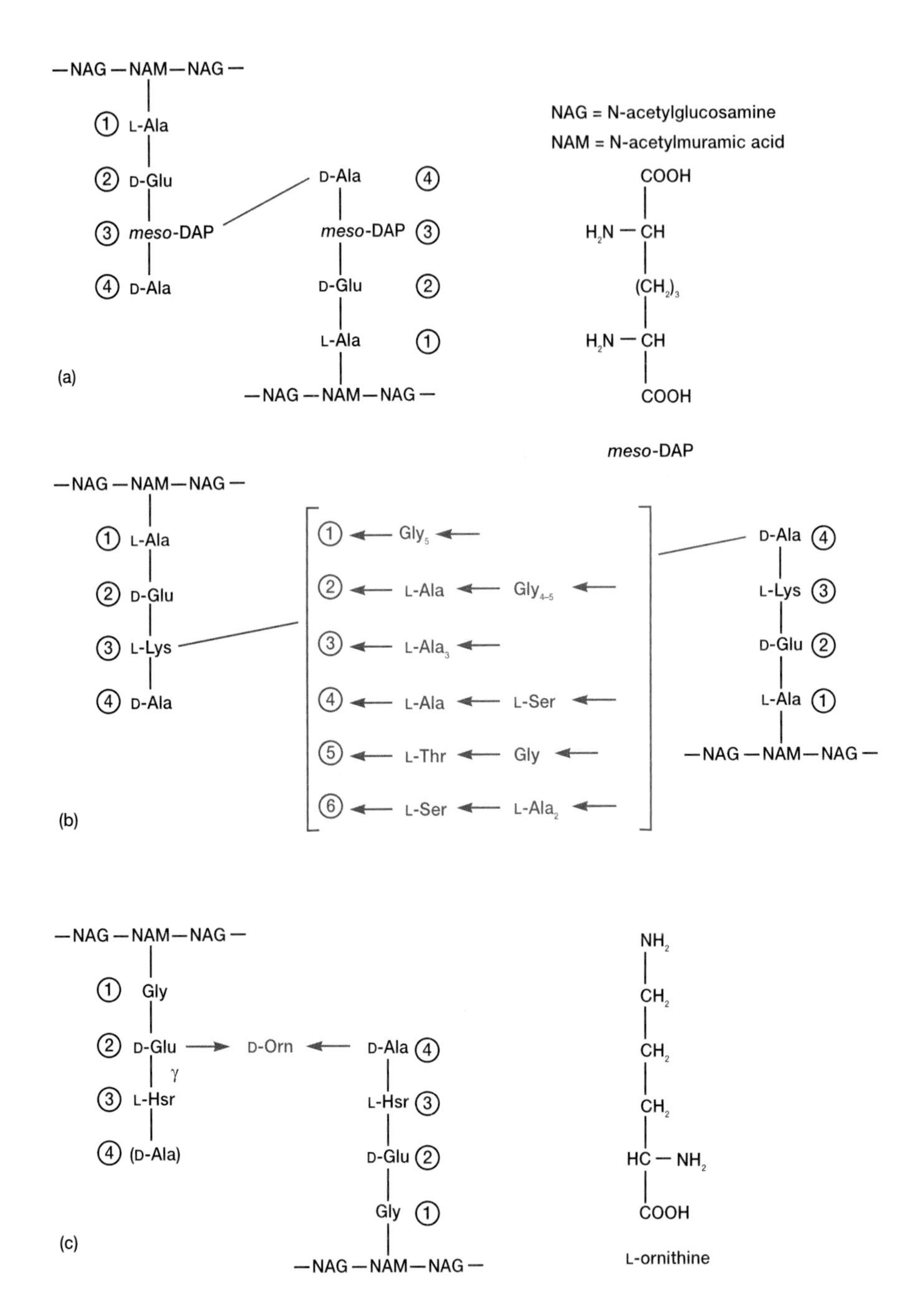

Representative Examples of Peptidoglycan Structure
Figure 21.1

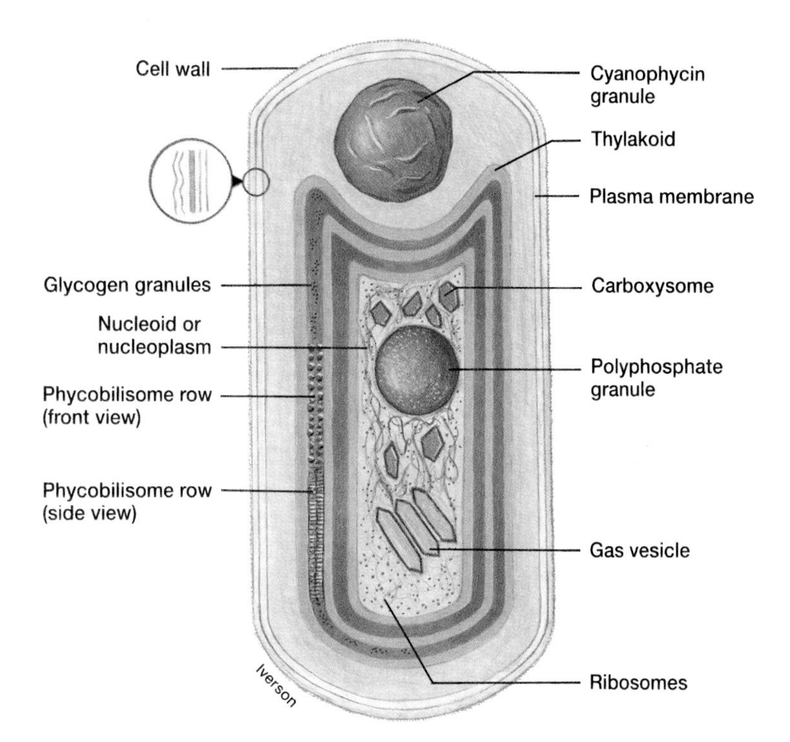

Cell wall

Cyanophycin granule

Thylakoid

Plasma membrane

Glycogen granules

Carboxysome

Nucleoid or nucleoplasm

Polyphosphate granule

Phycobilisome row (front view)

Phycobilisome row (side view)

Gas vesicle

Iverson

Ribosomes

Cyanobacterial Cell Structure
Figure 22.10 *a*

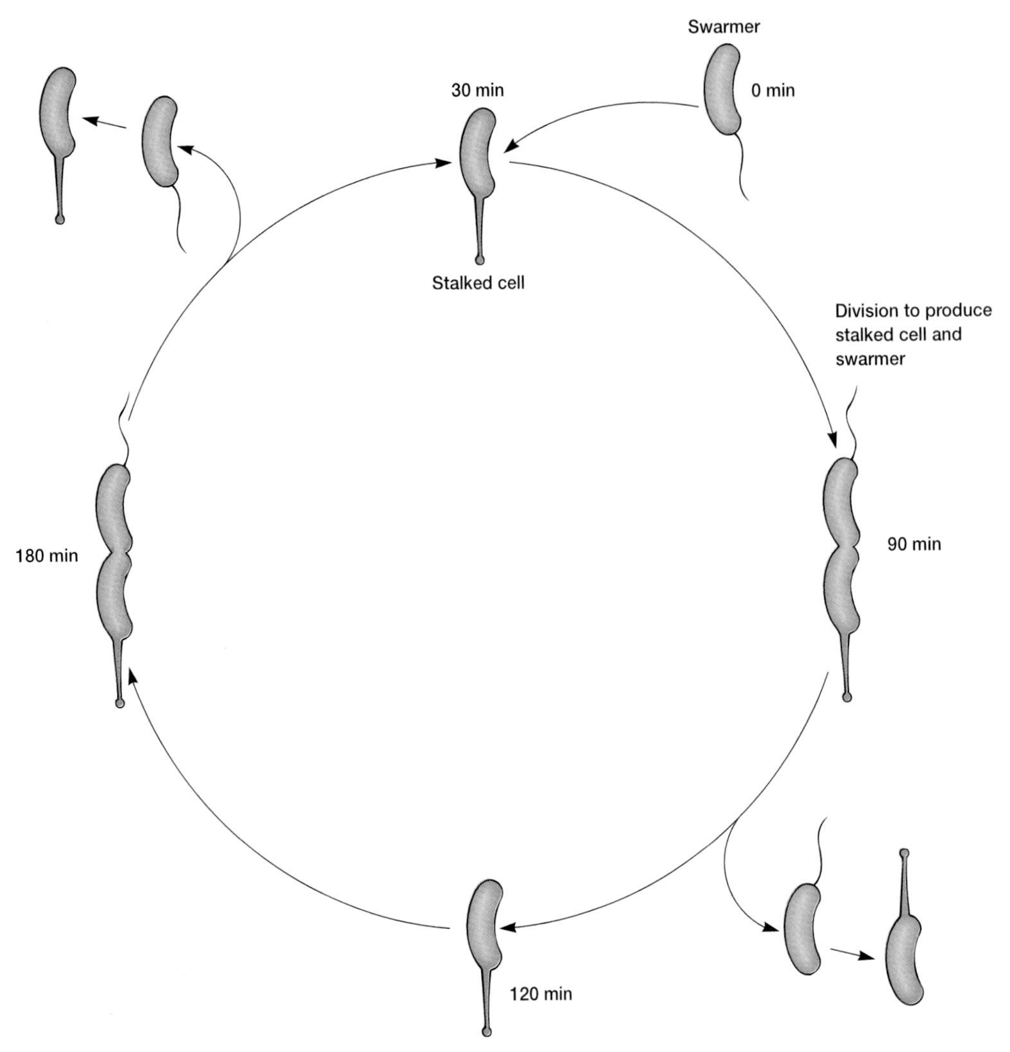

Swarmer

0 min

30 min

Stalked cell

Division to produce
stalked cell and
swarmer

90 min

180 min

120 min

Caulobacter Life Cycle
Figure 22.19

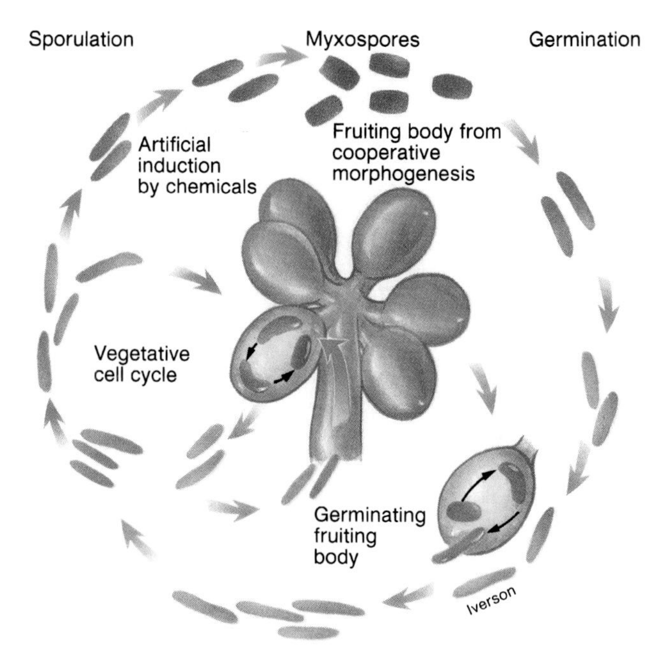

Myxobacterial Life Cycle
Figure 22.27

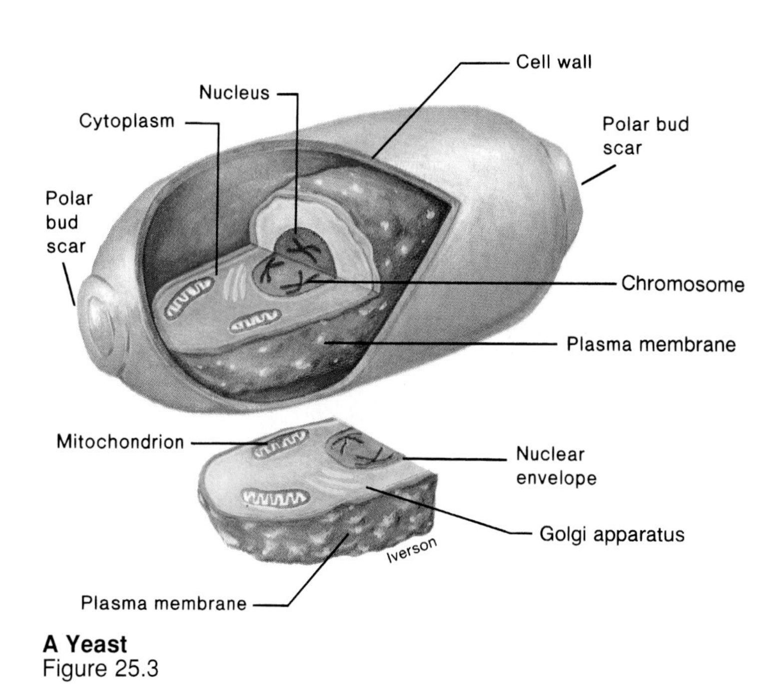

A Yeast
Figure 25.3

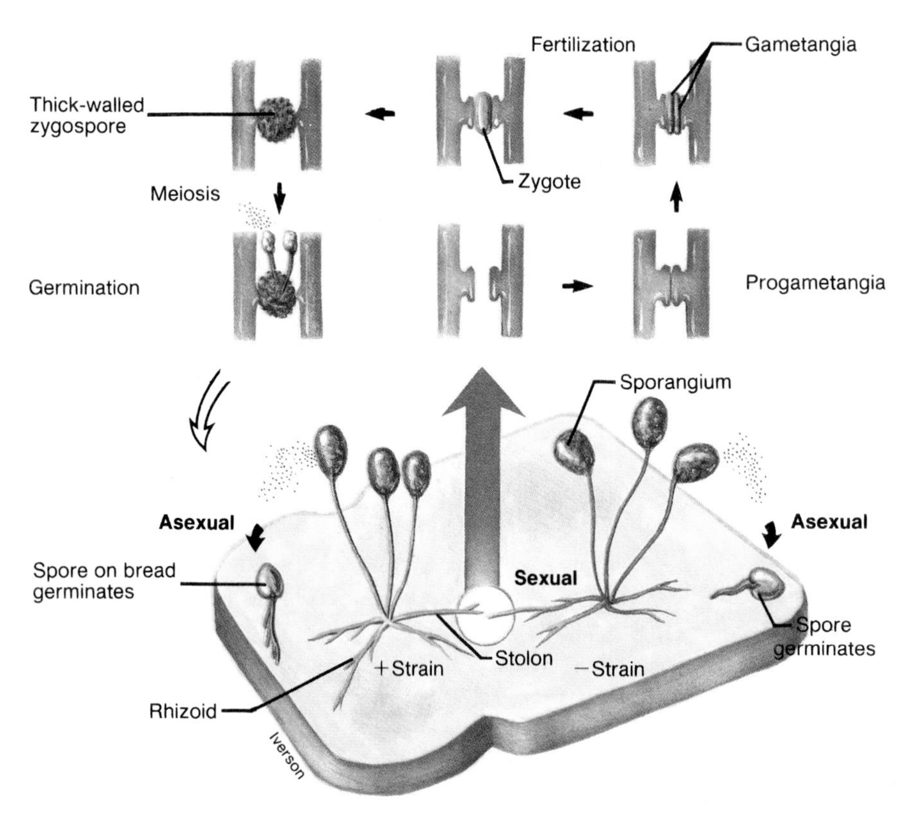

Fertilization

Gametangia

Thick-walled zygospore

Zygote

Meiosis

Germination

Progametangia

Sporangium

Asexual

Asexual

Spore on bread germinates

Sexual

Spore germinates

+ Strain

Stolon

— Strain

Rhizoid

Iverson

Life Cycle of *Rhizopus stolonifer*
Figure 25.9

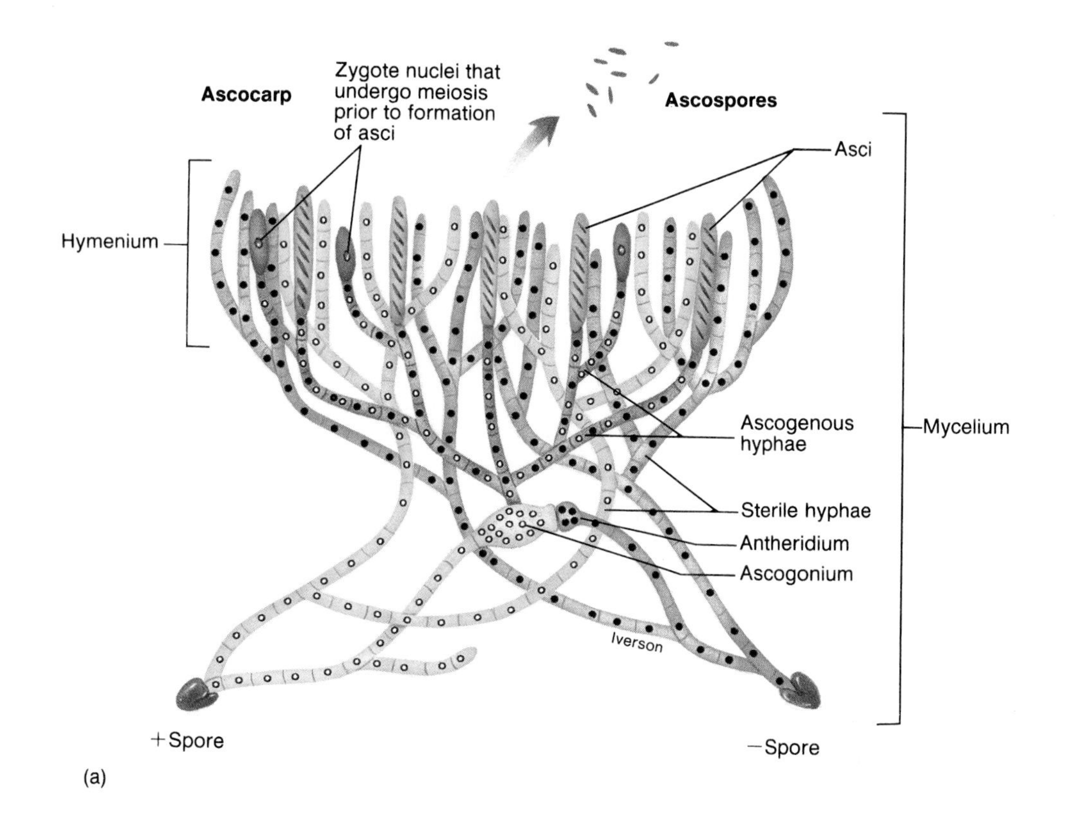

Ascocarp
Zygote nuclei that undergo meiosis prior to formation of asci
Ascospores
Asci
Hymenium
Ascogenous hyphae
Mycelium
Sterile hyphae
Antheridium
Ascogonium
Iverson
+Spore
−Spore
(a)

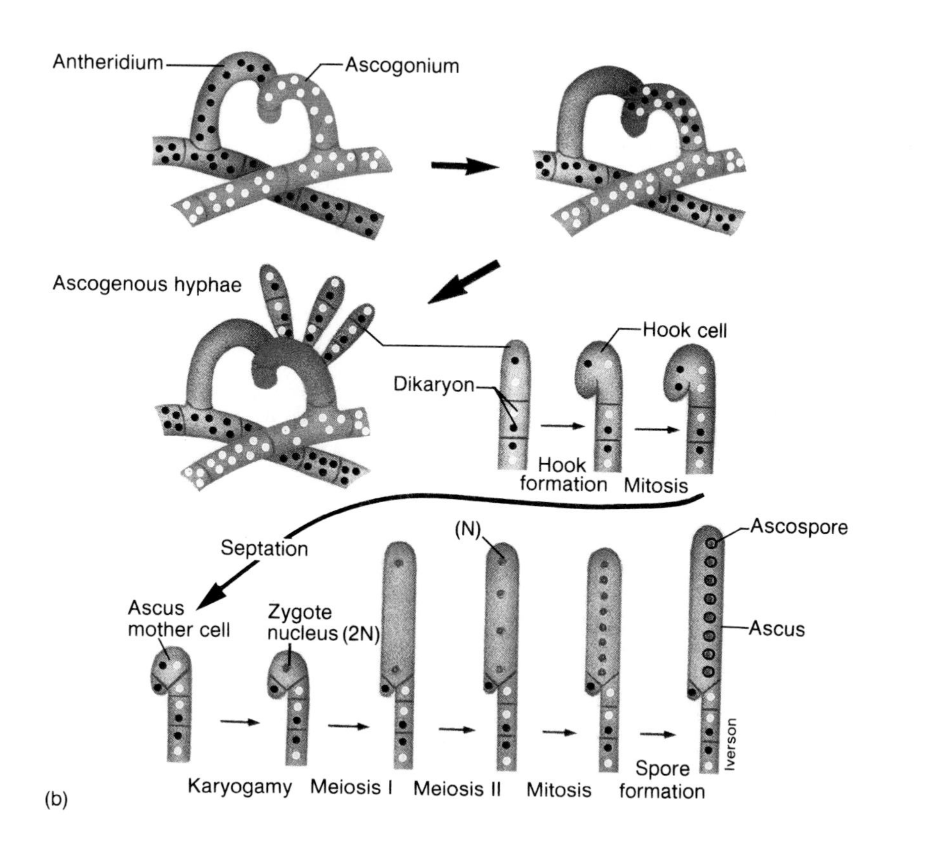

Antheridium
Ascogonium
Ascogenous hyphae
Hook cell
Dikaryon
Hook formation
Mitosis
Septation
(N)
Ascospore
Ascus
Ascus mother cell
Zygote nucleus (2N)
Karyogamy
Meiosis I
Meiosis II
Mitosis
Spore formation
Iverson
(b)

Life Cycle of Ascomycetes
Figure 25.12

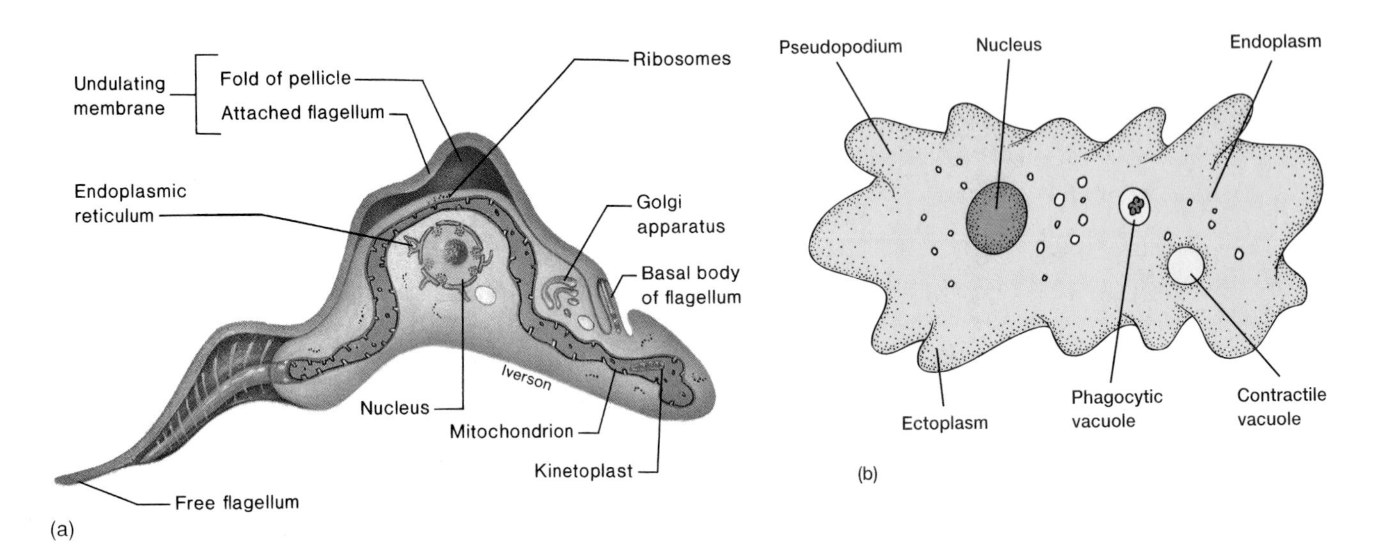

(a)

Undulating membrane
Fold of pellicle
Attached flagellum
Ribosomes
Endoplasmic reticulum
Golgi apparatus
Basal body of flagellum
Iverson
Nucleus
Mitochondrion
Kinetoplast
Free flagellum

(b)

Pseudopodium
Nucleus
Endoplasm
Ectoplasm
Phagocytic vacuole
Contractile vacuole

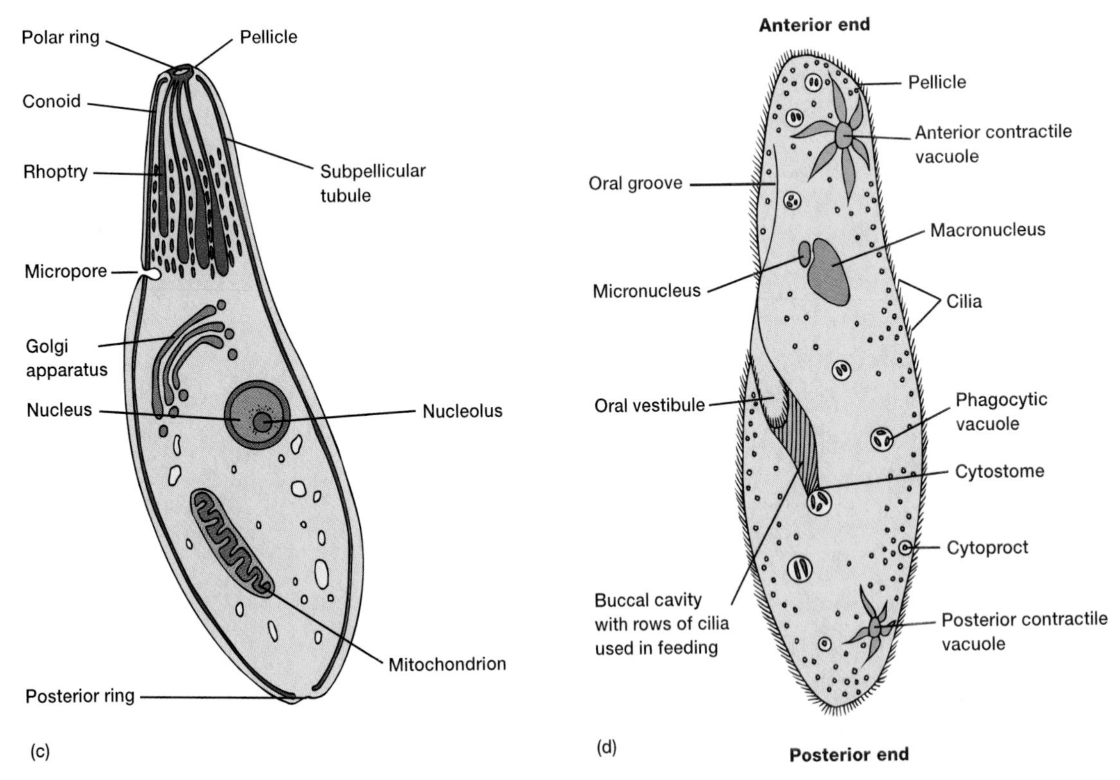

(c)

Polar ring
Pellicle
Conoid
Rhoptry
Subpellicular tubule
Micropore
Golgi apparatus
Nucleus
Nucleolus
Mitochondrion
Posterior ring

(d)

Anterior end

Pellicle
Anterior contractile vacuole
Oral groove
Macronucleus
Micronucleus
Cilia
Oral vestibule
Phagocytic vacuole
Cytostome
Cytoproct
Buccal cavity with rows of cilia used in feeding
Posterior contractile vacuole

Posterior end

Representative Protozoa
Figure 27.3

72

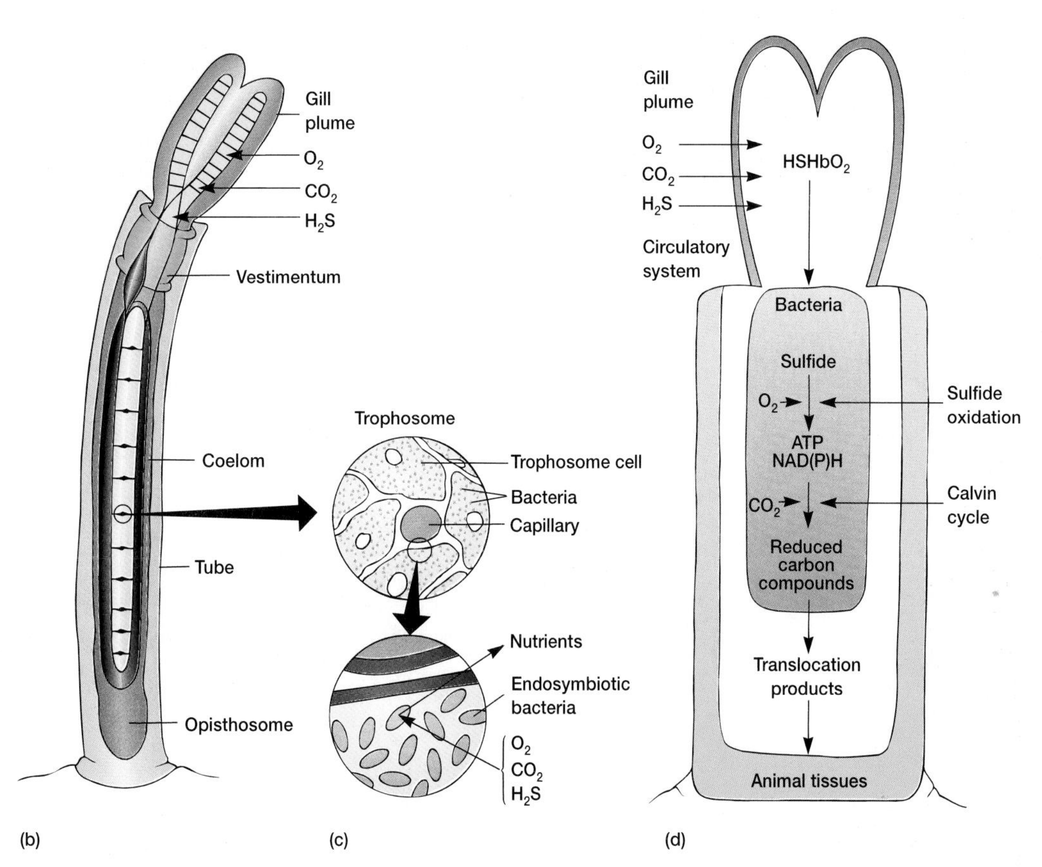

(b)

(c)

(d)

The Tube Worm–Bacterial Relationship
Figure 28.4 *b–d*

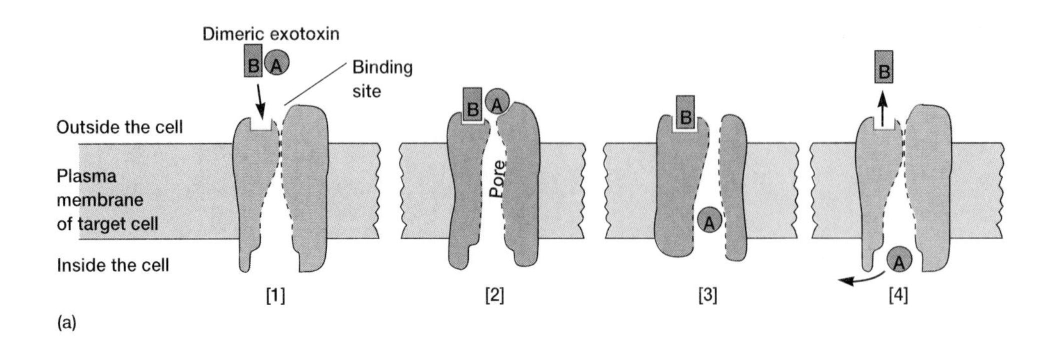
Dimeric exotoxin
B
A
Binding site
Outside the cell
Plasma membrane of target cell
Inside the cell
Pore
[1]
[2]
[3]
[4]
(a)

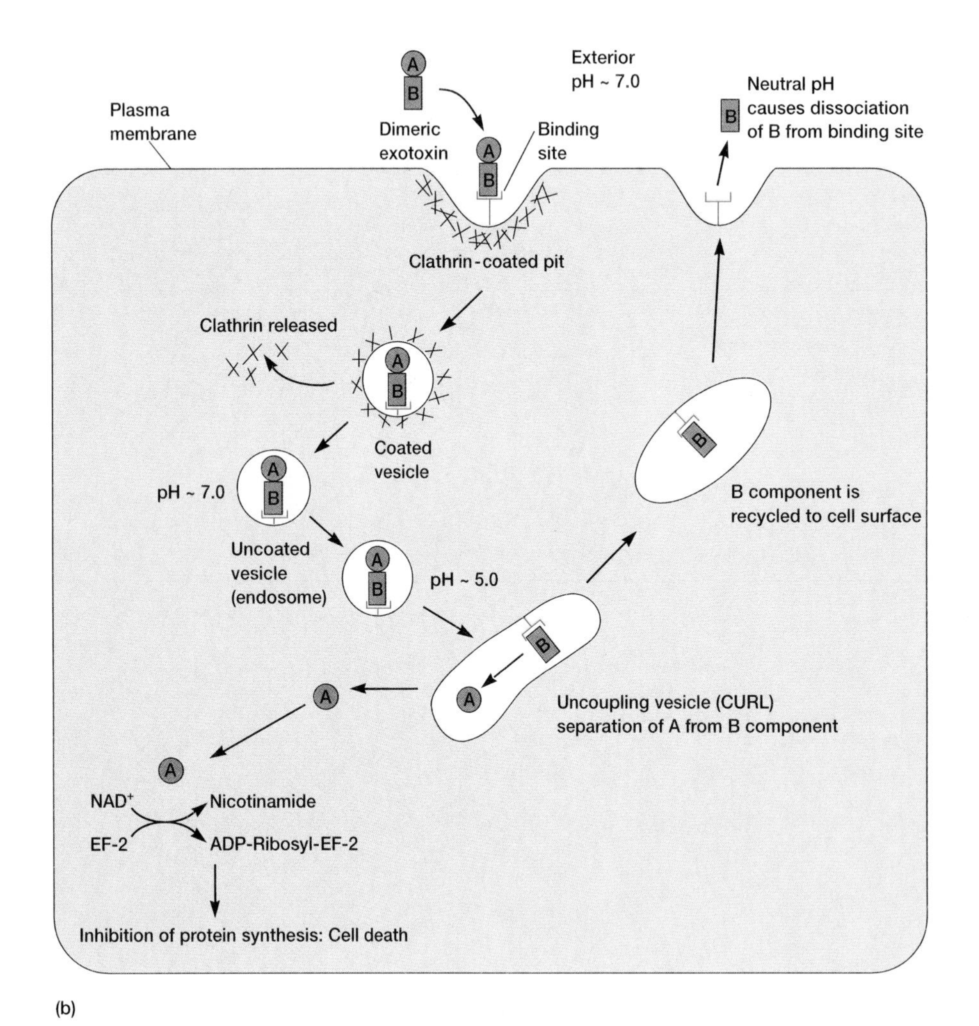
Plasma membrane
Exterior pH ~ 7.0
Dimeric exotoxin
Binding site
Neutral pH causes dissociation of B from binding site
Clathrin-coated pit
Clathrin released
Coated vesicle
pH ~ 7.0
Uncoated vesicle (endosome)
pH ~ 5.0
B component is recycled to cell surface
Uncoupling vesicle (CURL) separation of A from B component
NAD+
Nicotinamide
EF-2
ADP-Ribosyl-EF-2
Inhibition of protein synthesis: Cell death

(b)

Exotoxin Transport Mechanisms
Figure 29.4

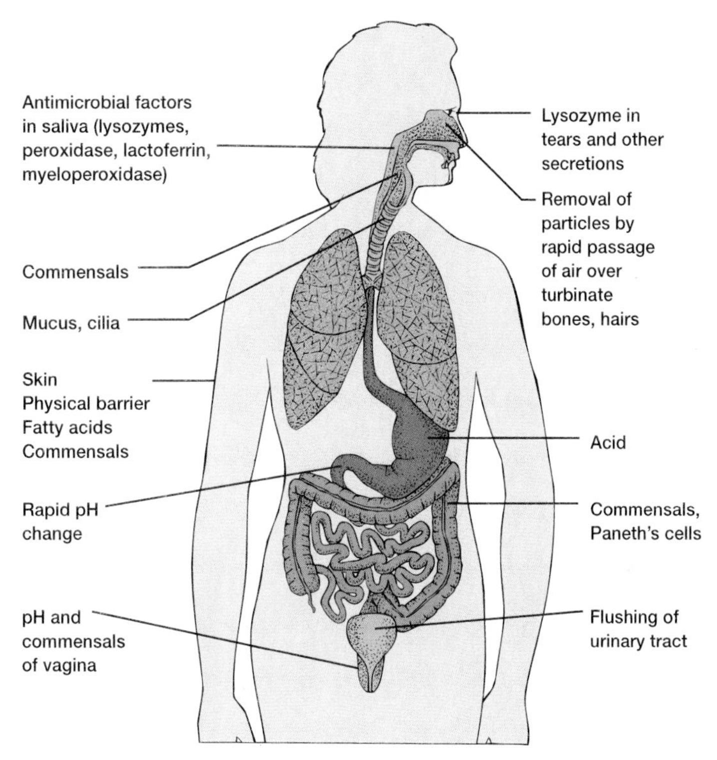

Antimicrobial factors
in saliva (lysozymes,
peroxidase, lactoferrin,
myeloperoxidase)

Lysozyme in
tears and other
secretions

Removal of
particles by
rapid passage
of air over
turbinate
bones, hairs

Commensals

Mucus, cilia

Skin
Physical barrier
Fatty acids
Commensals

Acid

Commensals,
Paneth's cells

Rapid pH
change

pH and
commensals
of vagina

Flushing of
urinary tract

Host Defenses
Figure 29.6

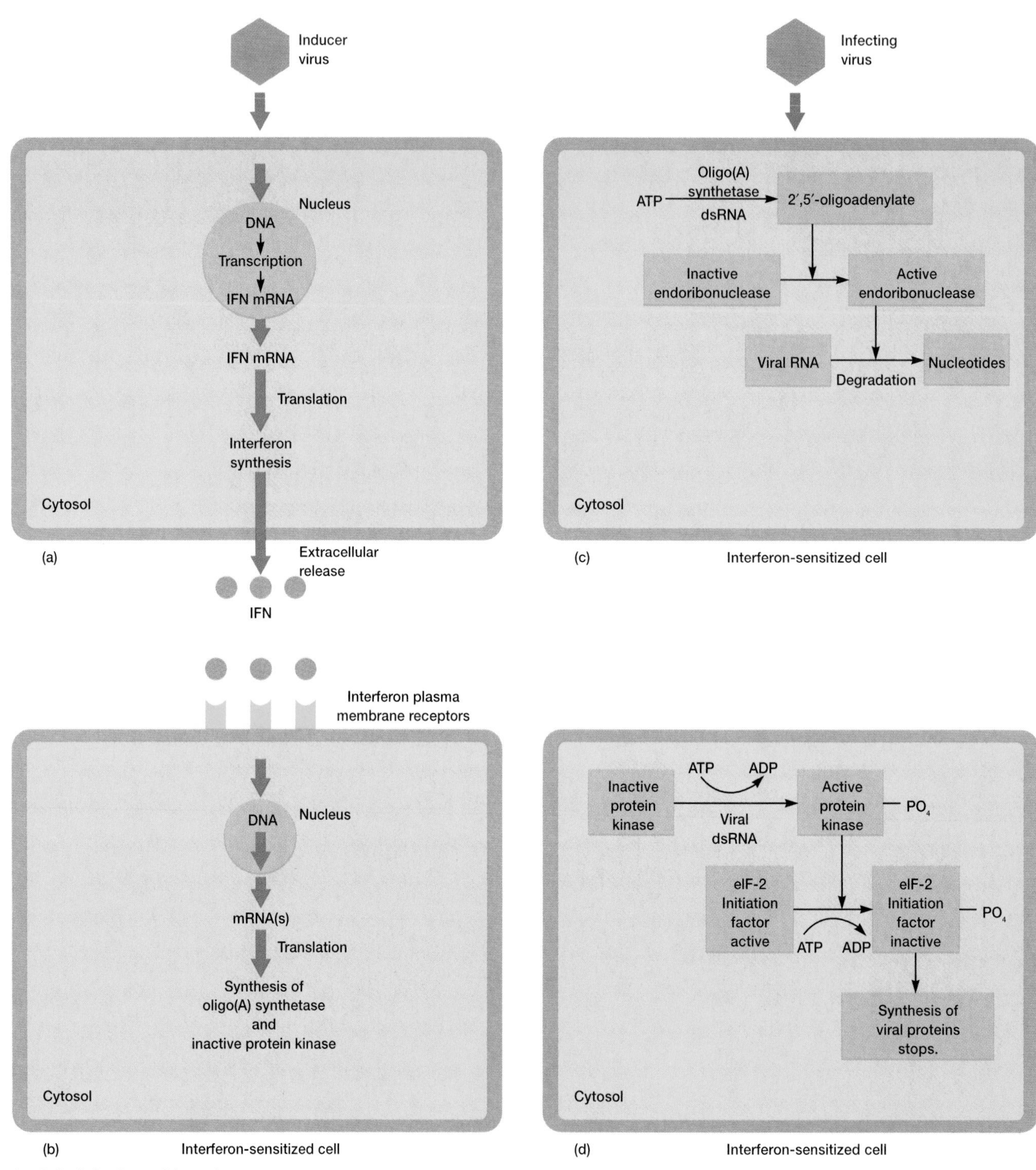

Inducer virus
Nucleus
DNA
Transcription
IFN mRNA
IFN mRNA
Translation
Interferon synthesis
Cytosol
(a)
Extracellular release
IFN
Interferon plasma membrane receptors
DNA
Nucleus
mRNA(s)
Translation
Synthesis of oligo(A) synthetase and inactive protein kinase
Cytosol
(b)
Interferon-sensitized cell
Infecting virus
Oligo(A) synthetase
ATP
dsRNA
2′,5′-oligoadenylate
Inactive endoribonuclease
Active endoribonuclease
Viral RNA
Degradation
Nucleotides
Cytosol
(c)
Interferon-sensitized cell
ATP
ADP
Inactive protein kinase
Viral dsRNA
Active protein kinase
PO4
eIF-2 Initiation factor active
ATP
ADP
eIF-2 Initiation factor inactive
PO4
Synthesis of viral proteins stops.
Cytosol
(d)
Interferon-sensitized cell

Antiviral Action of Interferon
Figure 29.9

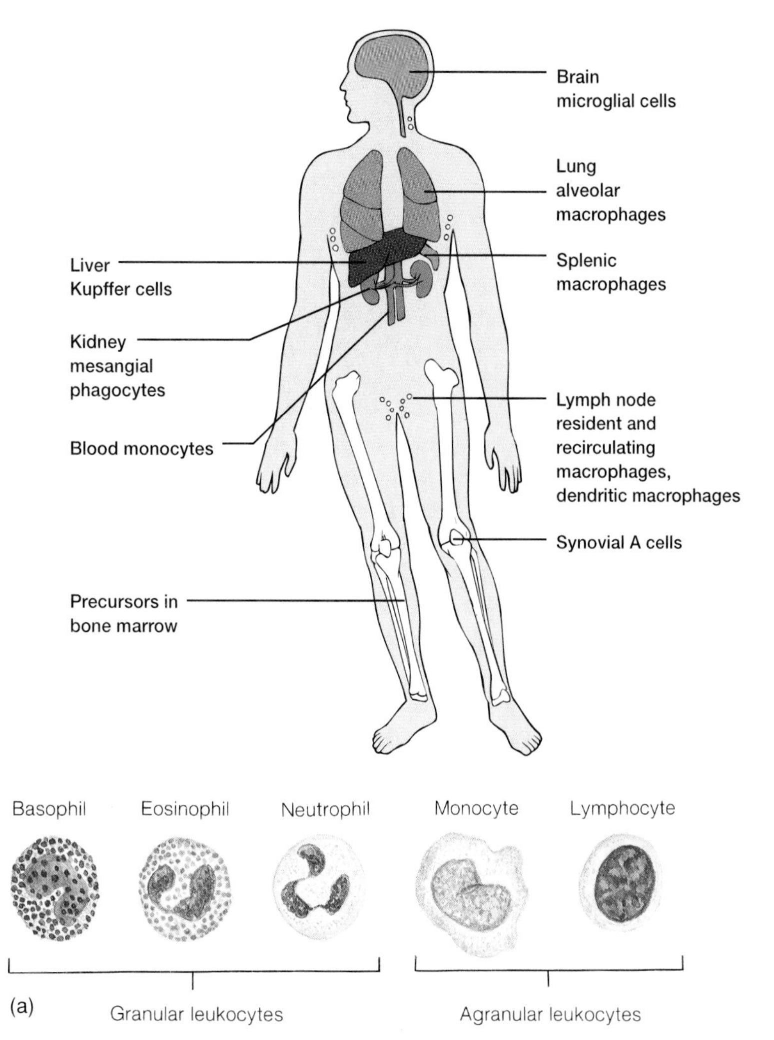

Brain
microglial cells

Lung
alveolar
macrophages

Liver
Kupffer cells

Splenic
macrophages

Kidney
mesangial
phagocytes

Blood monocytes

Lymph node
resident and
recirculating
macrophages,
dendritic macrophages

Synovial A cells

Precursors in
bone marrow

Basophil   Eosinophil   Neutrophil   Monocyte   Lymphocyte

(a)   Granular leukocytes   Agranular leukocytes

The RE System and Cells of the Immune System
Figure 29.10 & 29.11 *a*

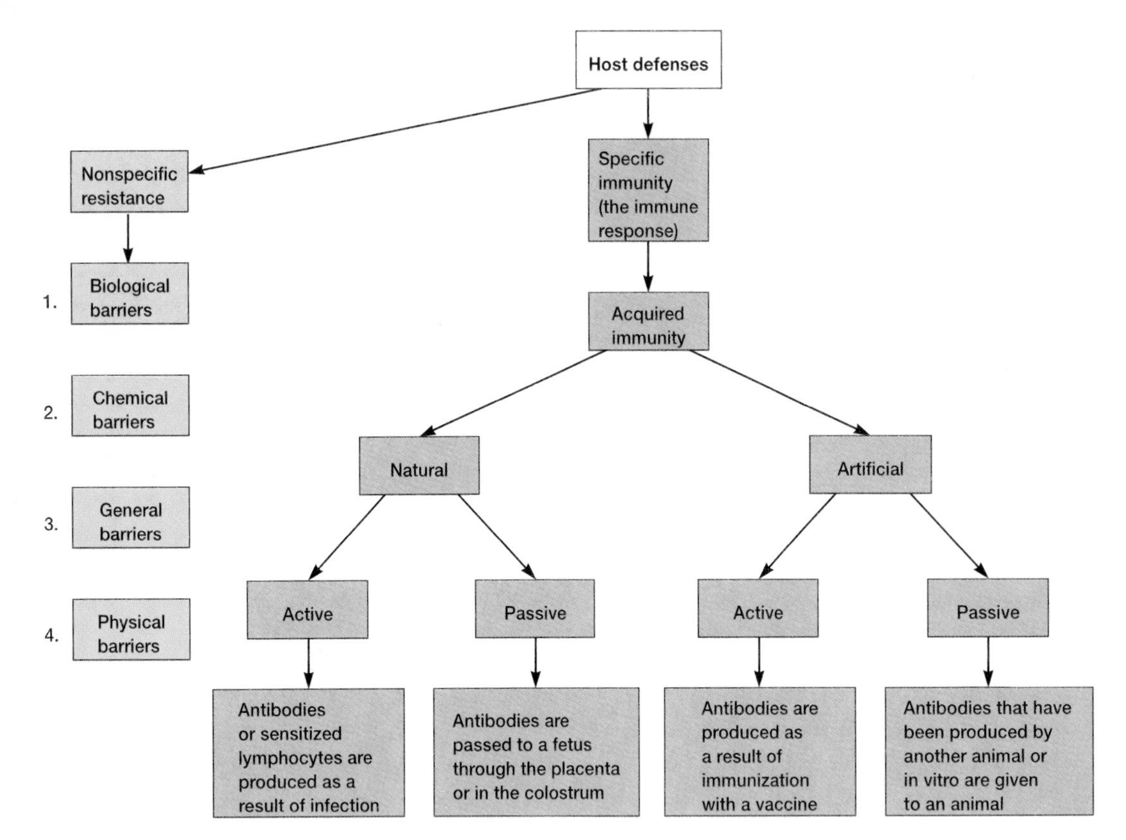

Different Host Defenses
Figure 30.1

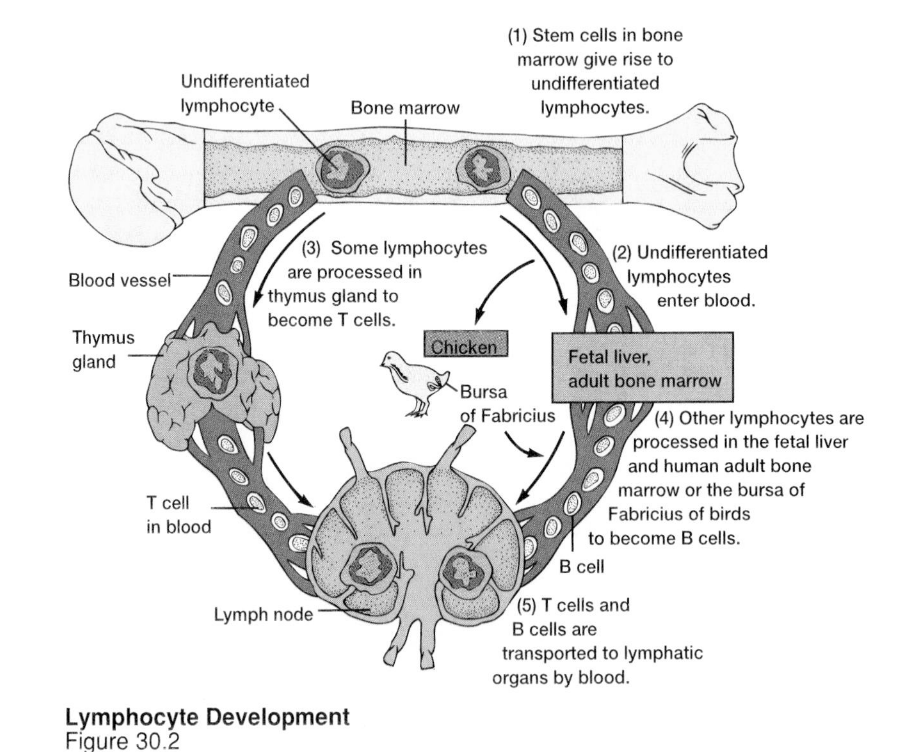

Lymphocyte Development
Figure 30.2

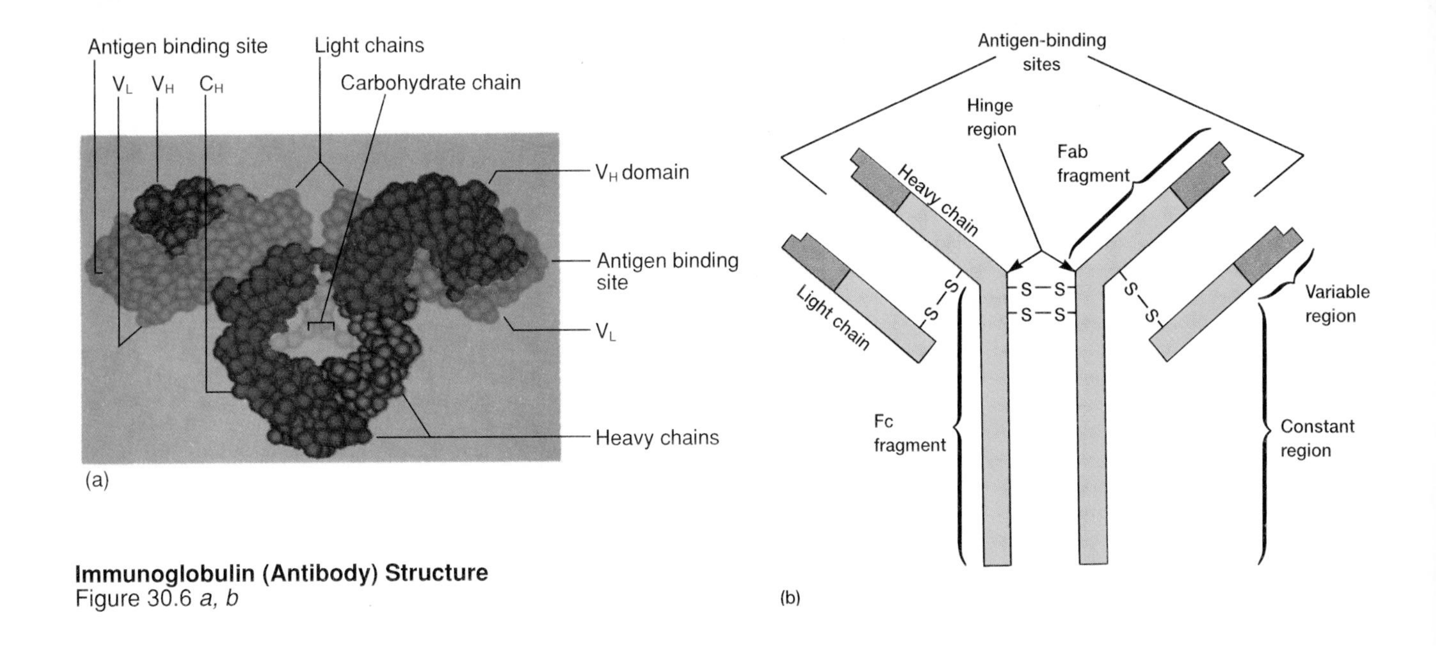

Immunoglobulin (Antibody) Structure
Figure 30.6 *a, b*

(a) Labels: Antigen binding site, V_L, V_H, C_H, Light chains, Carbohydrate chain, V_H domain, Antigen binding site, V_L, Heavy chains

(b) Labels: Antigen-binding sites, Hinge region, Fab fragment, Heavy chain, Light chain, Variable region, Fc fragment, Constant region

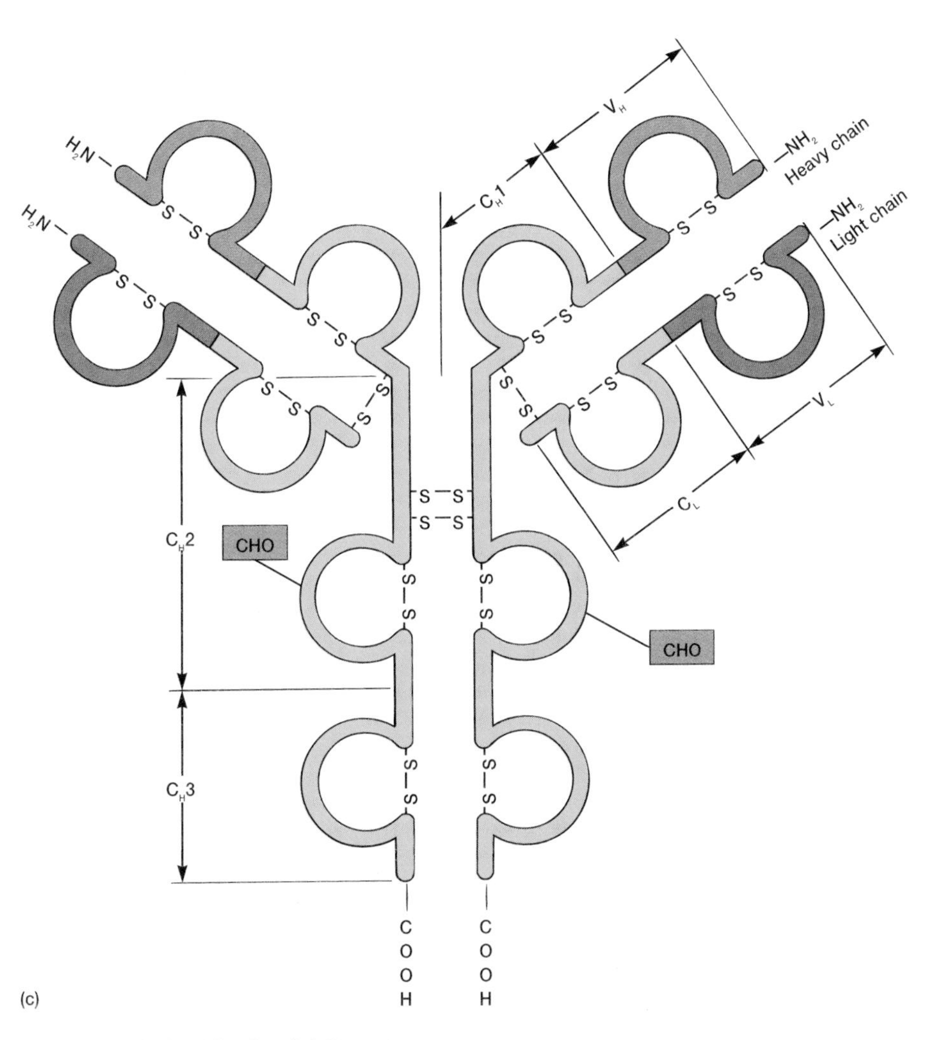

Immunoglobulin (Antibody) Structure
Figure 30.6 *c*

(c) Labels: H_2N, V_H, C_H1, NH_2 Heavy chain, NH_2 Light chain, V_L, C_L, C_H2, CHO, CHO, C_H3, COOH, COOH

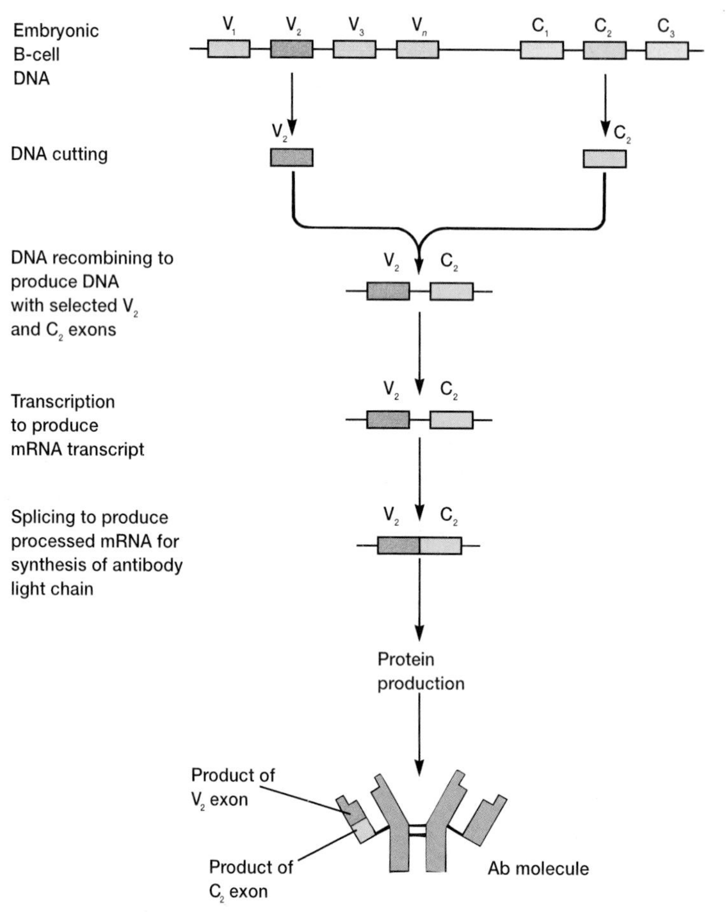

Gene Shuffling and Antibody Diversity
Figure 30.14

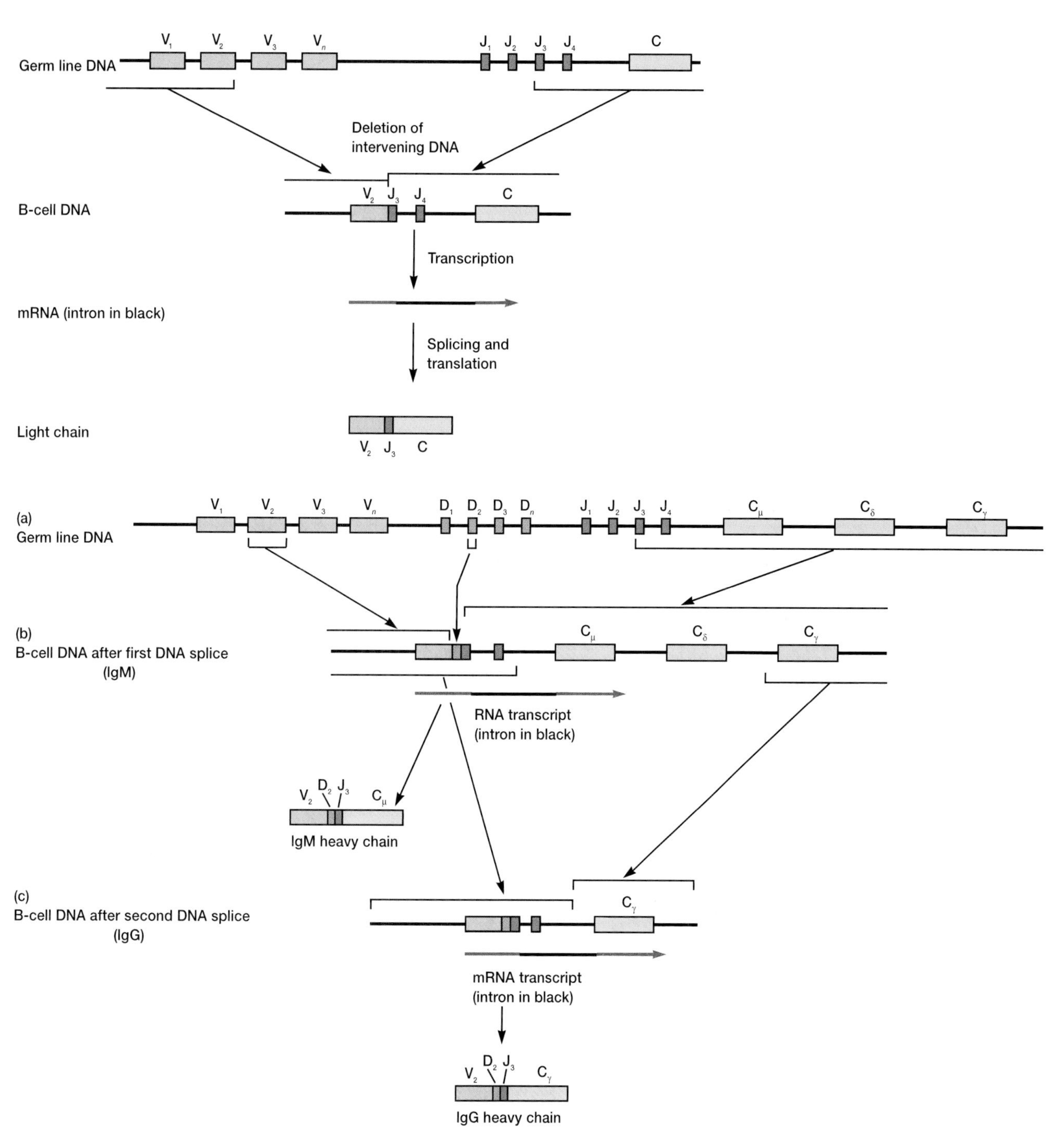

Light and Heavy Chain Production
Figure 30.15 & 30.16

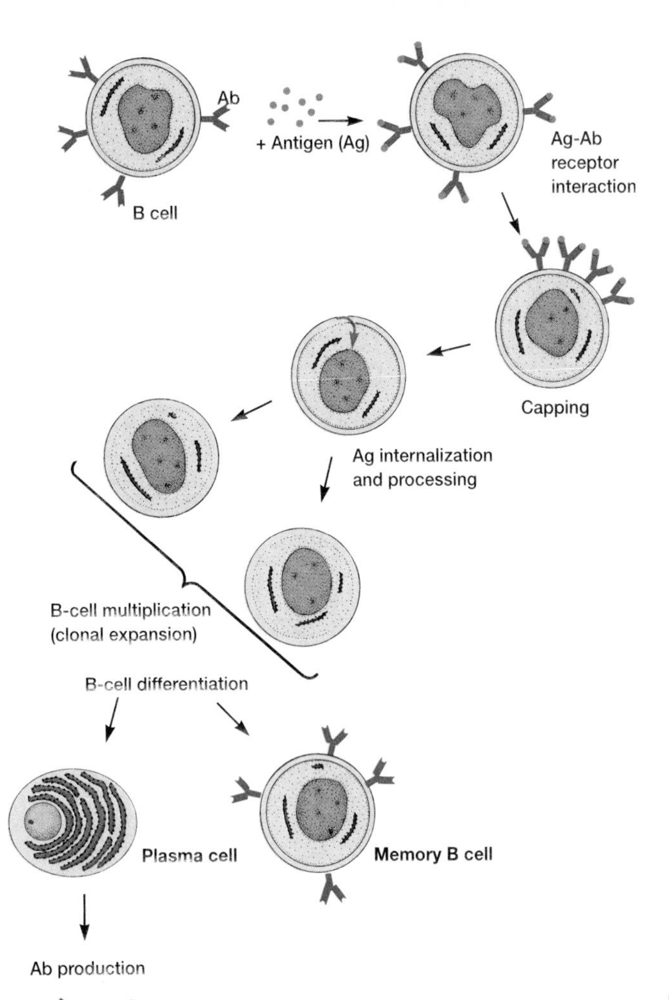

Clonal Selection
Figure 30.17

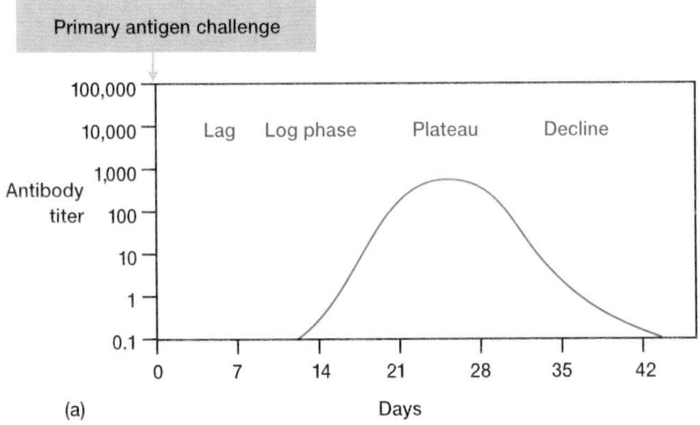

Primary antigen challenge

Lag   Log phase   Plateau   Decline

Antibody titer

Days

(a)

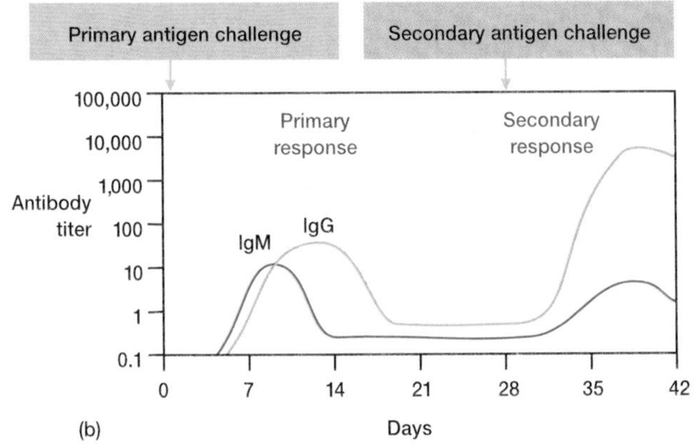

Primary antigen challenge

Secondary antigen challenge

Primary response

Secondary response

IgM   IgG

Antibody titer

Days

(b)

Primary & Secondary Antibody Responses
Figure 30.18

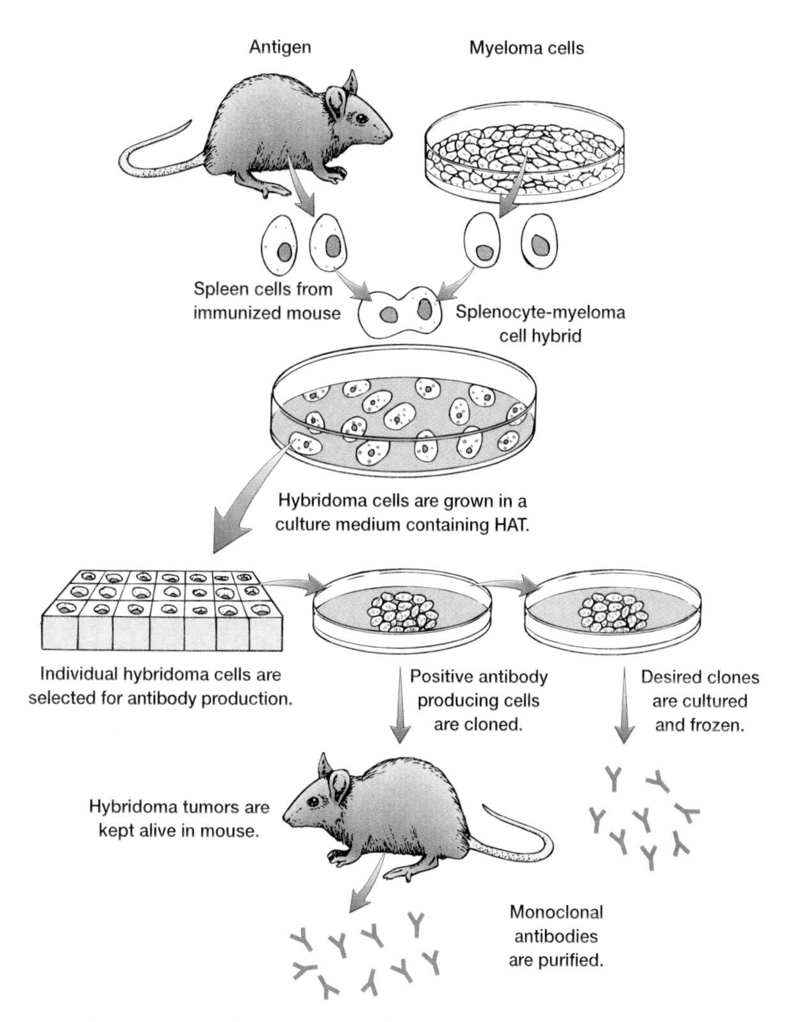

Antigen

Myeloma cells

Spleen cells from
immunized mouse

Splenocyte-myeloma
cell hybrid

Hybridoma cells are grown in a
culture medium containing HAT.

Individual hybridoma cells are
selected for antibody production.

Positive antibody
producing cells
are cloned.

Desired clones
are cultured
and frozen.

Hybridoma tumors are
kept alive in mouse.

Monoclonal
antibodies
are purified.

The Production of Monoclonal Antibodies
Figure 30.19

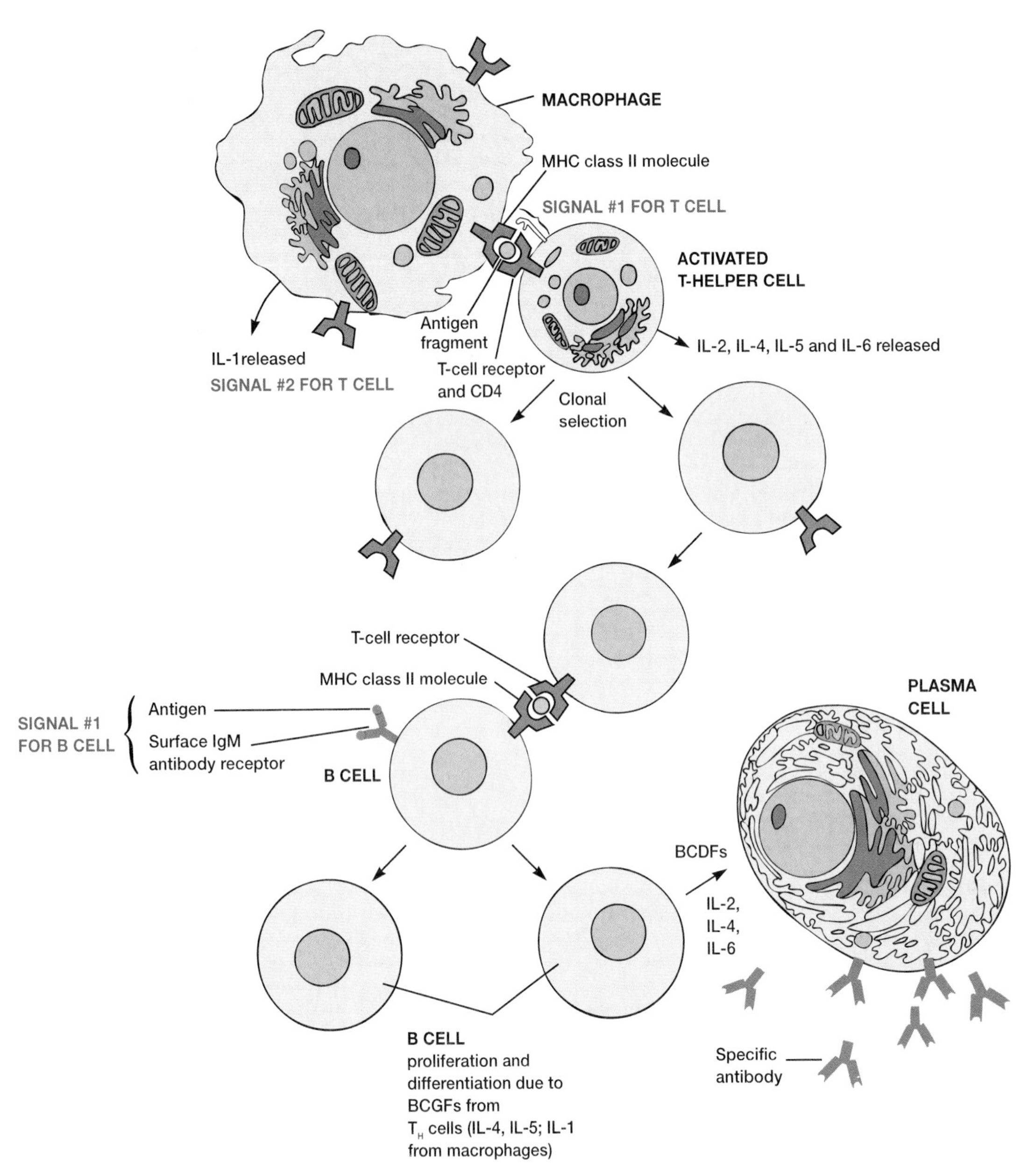

MACROPHAGE

MHC class II molecule

SIGNAL #1 FOR T CELL

ACTIVATED T-HELPER CELL

IL-1 released
SIGNAL #2 FOR T CELL

Antigen fragment

T-cell receptor and CD4

IL-2, IL-4, IL-5 and IL-6 released

Clonal selection

T-cell receptor

MHC class II molecule

SIGNAL #1 FOR B CELL { Antigen
Surface IgM antibody receptor

B CELL

PLASMA CELL

BCDFs

IL-2, IL-4, IL-6

B CELL
proliferation and differentiation due to BCGFs from T$_H$ cells (IL-4, IL-5; IL-1 from macrophages)

Specific antibody

T-Dependent Antigen Triggering of a B Cell
Figure 31.2

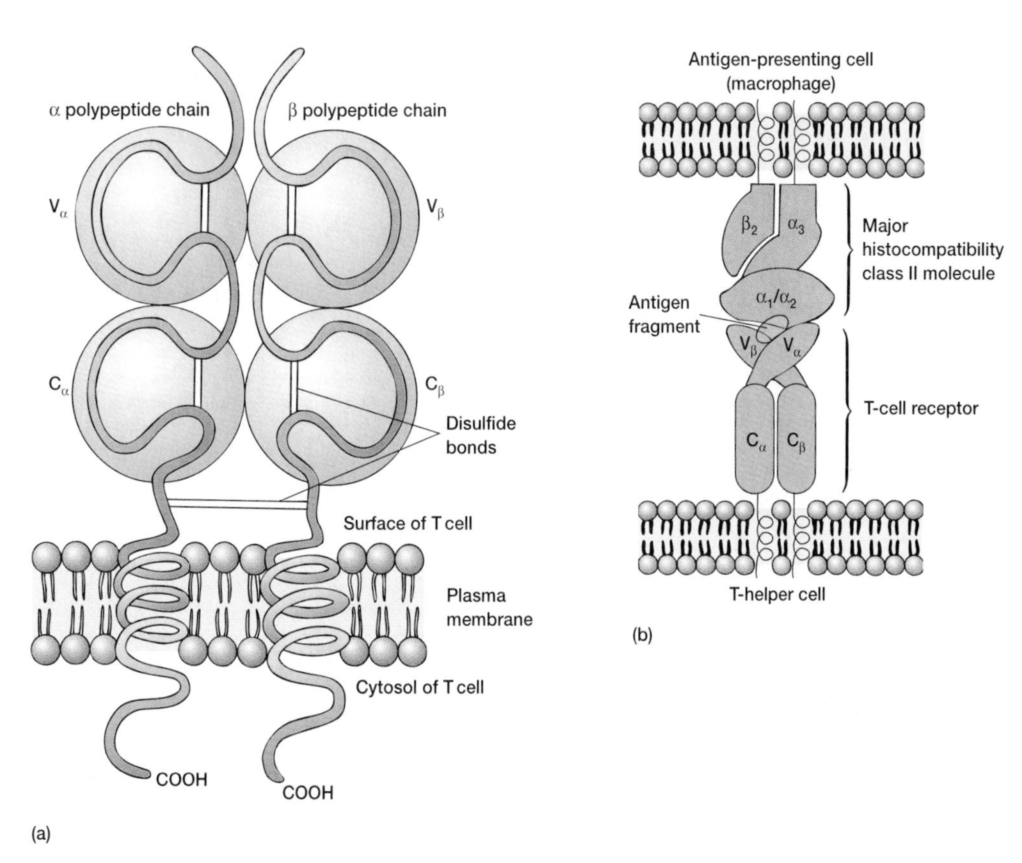

α polypeptide chain

β polypeptide chain

V_α

V_β

C_α

C_β

Disulfide bonds

Surface of T cell

Plasma membrane

Cytosol of T cell

COOH

COOH

(a)

Antigen-presenting cell (macrophage)

β_2 α_3

Major histocompatibility class II molecule

Antigen fragment

α_1/α_2

V_β V_α

C_α C_β

T-cell receptor

T-helper cell

(b)

T-Cell CD4 Receptor Protein and Activation
Figure 31.3

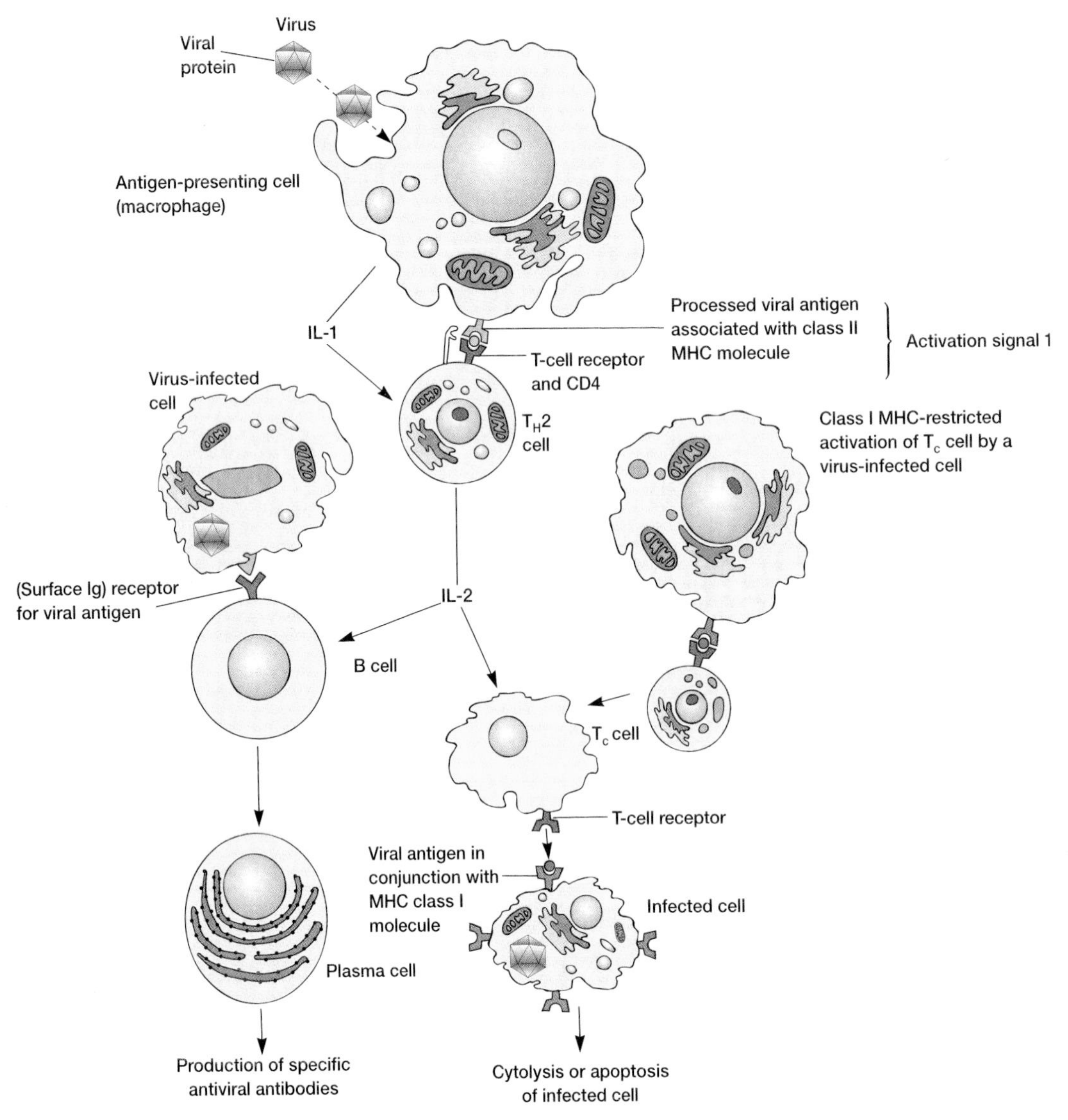

Virus

Viral
protein

Antigen-presenting cell
(macrophage)

IL-1

Virus-infected
cell

Processed viral antigen
associated with class II
MHC molecule

Activation signal 1

T-cell receptor
and CD4

T_H2
cell

Class I MHC-restricted
activation of T_c cell by a
virus-infected cell

(Surface Ig) receptor
for viral antigen

IL-2

B cell

T_c cell

T-cell receptor

Viral antigen in
conjunction with
MHC class I
molecule

Infected cell

Plasma cell

Production of specific
antiviral antibodies

Cytolysis or apoptosis
of infected cell

Regulator and Effector T Cells
Figure 31.6

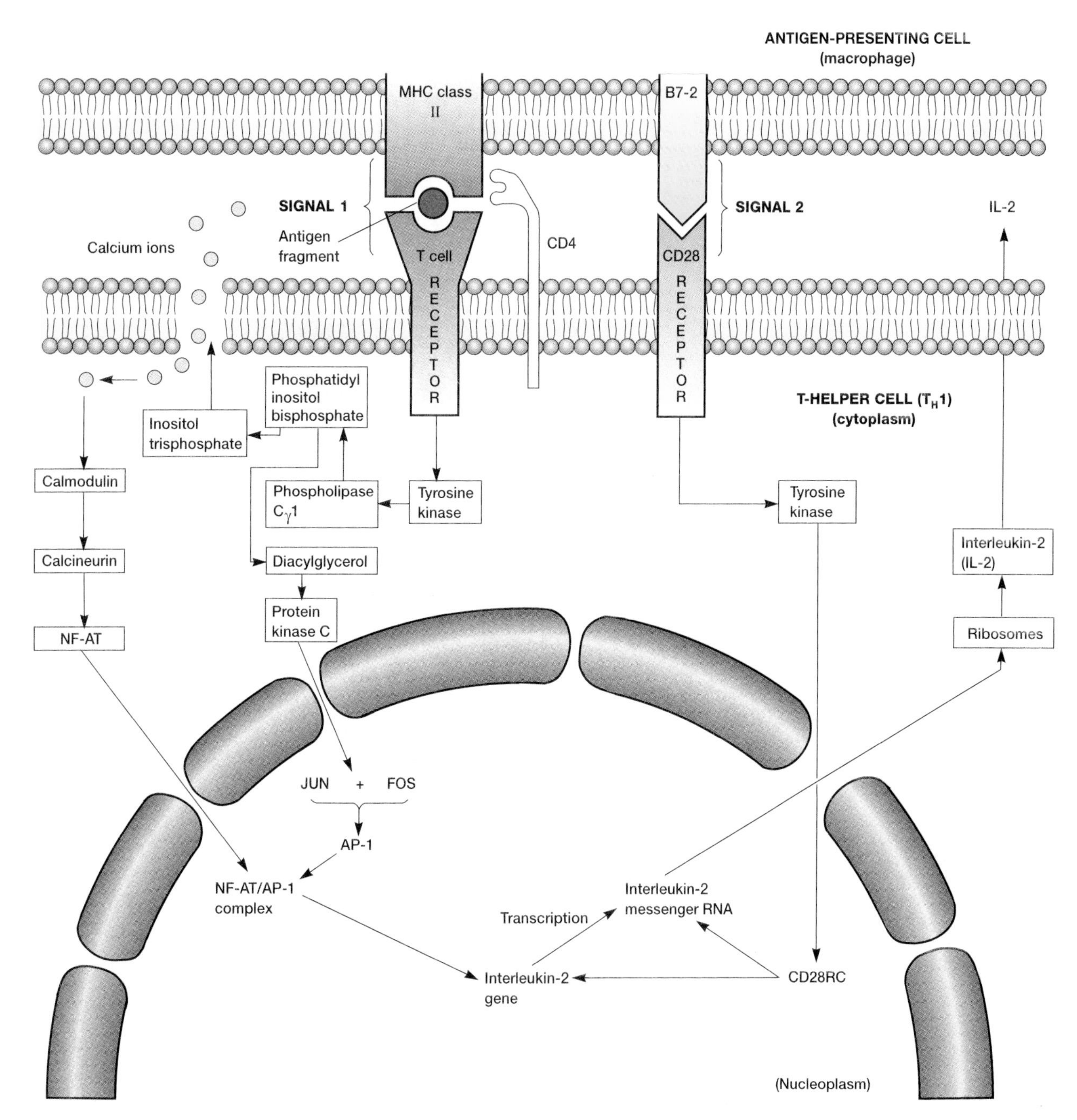

ANTIGEN-PRESENTING CELL
(macrophage)
MHC class II
B7-2
SIGNAL 1
SIGNAL 2
IL-2
Calcium ions
Antigen
fragment
T cell
RECEPTOR
CD4
CD28
RECEPTOR
Phosphatidyl
inositol
bisphosphate
Inositol
trisphosphate
T-HELPER CELL (T_H1)
(cytoplasm)
Calmodulin
Phospholipase
Cγ1
Tyrosine
kinase
Tyrosine
kinase
Interleukin-2
(IL-2)
Calcineurin
Diacylglycerol
Protein
kinase C
NF-AT
Ribosomes
JUN
+
FOS
AP-1
NF-AT/AP-1
complex
Interleukin-2
messenger RNA
Transcription
Interleukin-2
gene
CD28RC
(Nucleoplasm)

T-helper Cell Activation
Figure 31.7

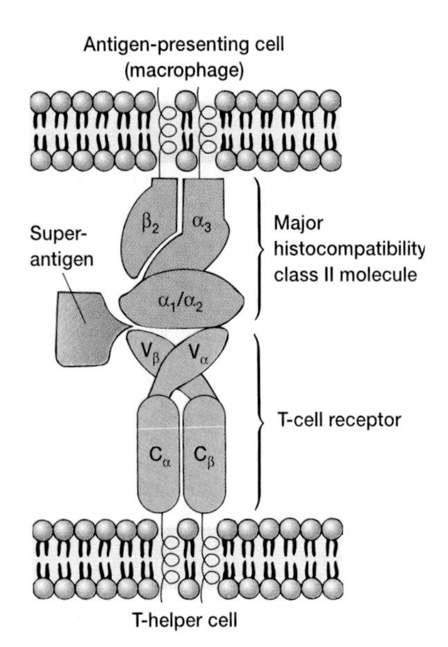

Antigen-presenting cell (macrophage)

Super-antigen

β_2 α_3

Major histocompatibility class II molecule

α_1/α_2

V_β V_α

T-cell receptor

C_α C_β

T-helper cell

Superantigens
Figure 31.9

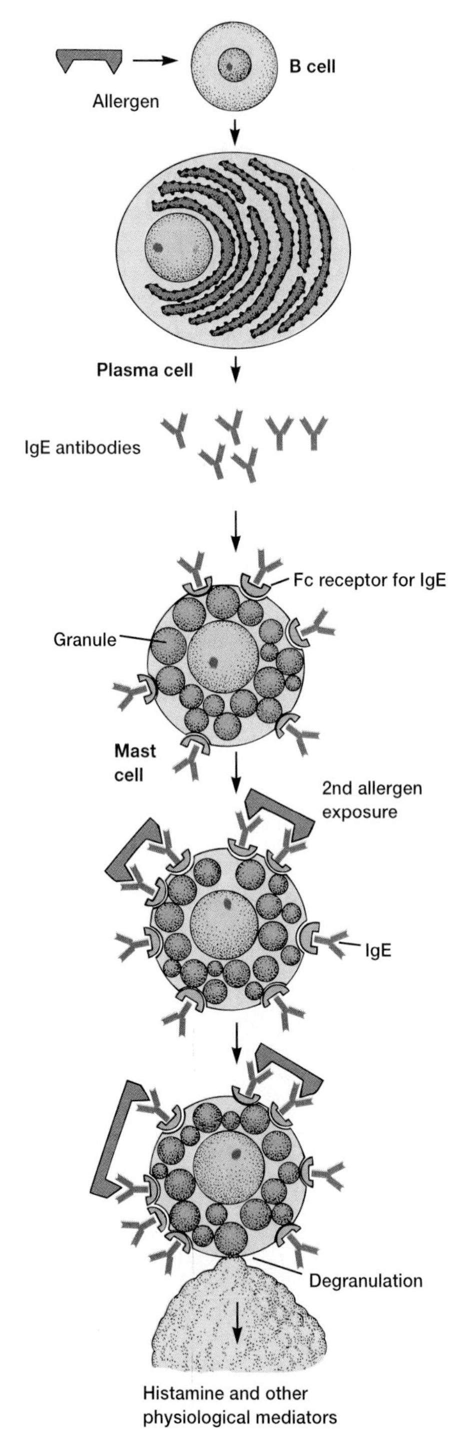

Allergen → B cell

Plasma cell

IgE antibodies

Fc receptor for IgE

Granule

Mast cell

2nd allergen exposure

IgE

Degranulation

Histamine and other physiological mediators

Type I (Anaphylaxis) Hypersensitivity
Figure 31.10

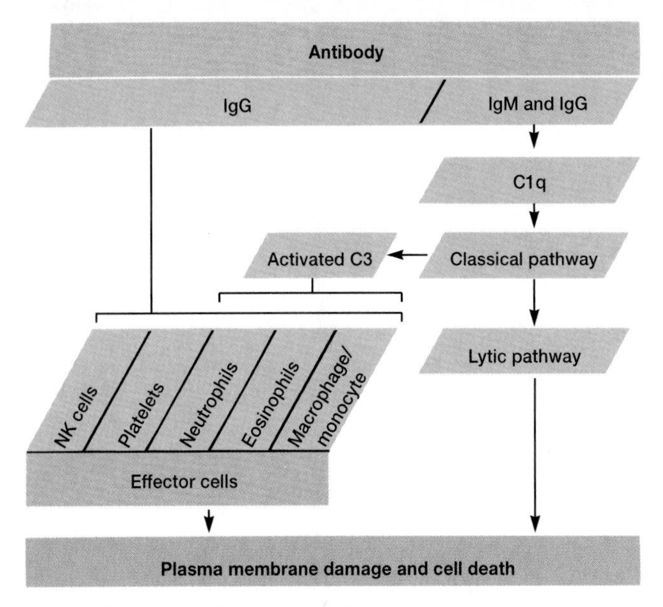

Type II (Cytotoxic) Hypersensitivity
Figure 31.12

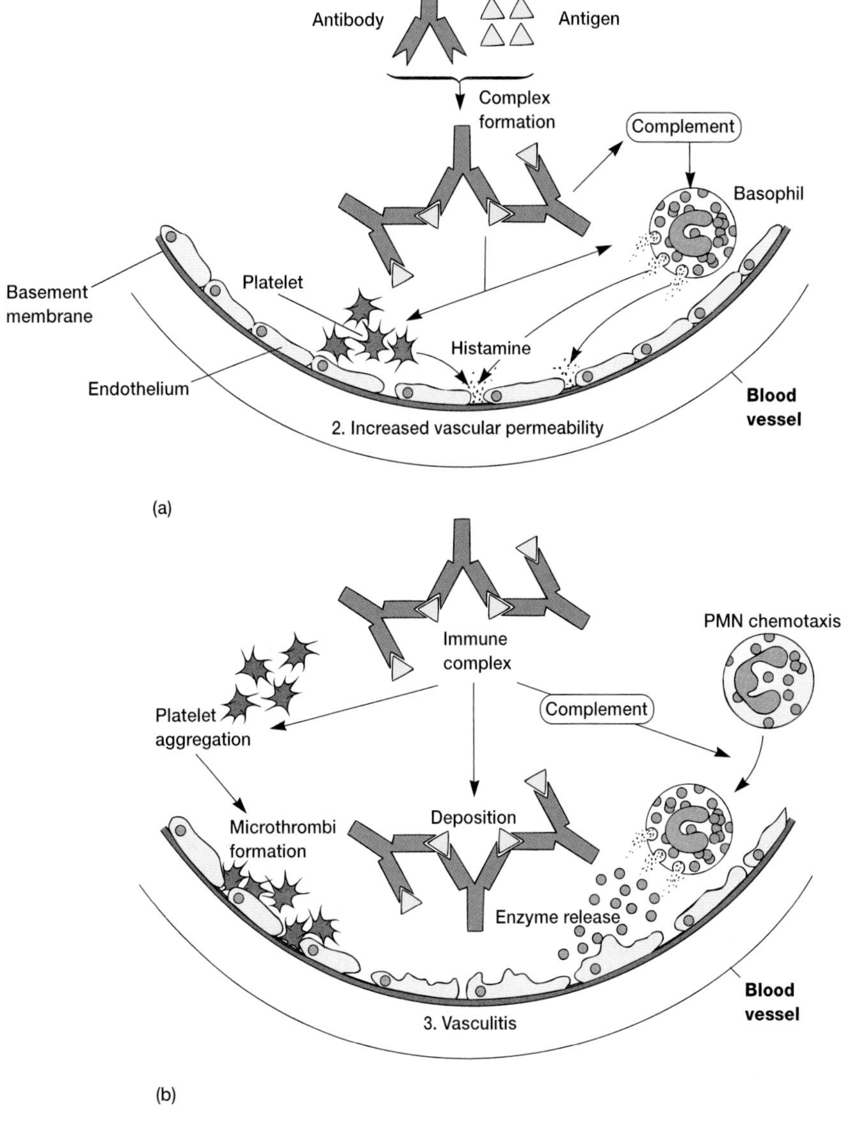

1. Complex formation

Antibody △△ Antigen
△△

Complex
formation

Complement

Basophil

Basement
membrane

Platelet

Histamine

Endothelium

**Blood
vessel**

2. Increased vascular permeability

(a)

Immune
complex

PMN chemotaxis

Complement

Platelet
aggregation

Microthrombi
formation

Deposition

Enzyme release

**Blood
vessel**

3. Vasculitis

(b)

Type III (Immune Complex) Hypersensitivity
Figure 31.13

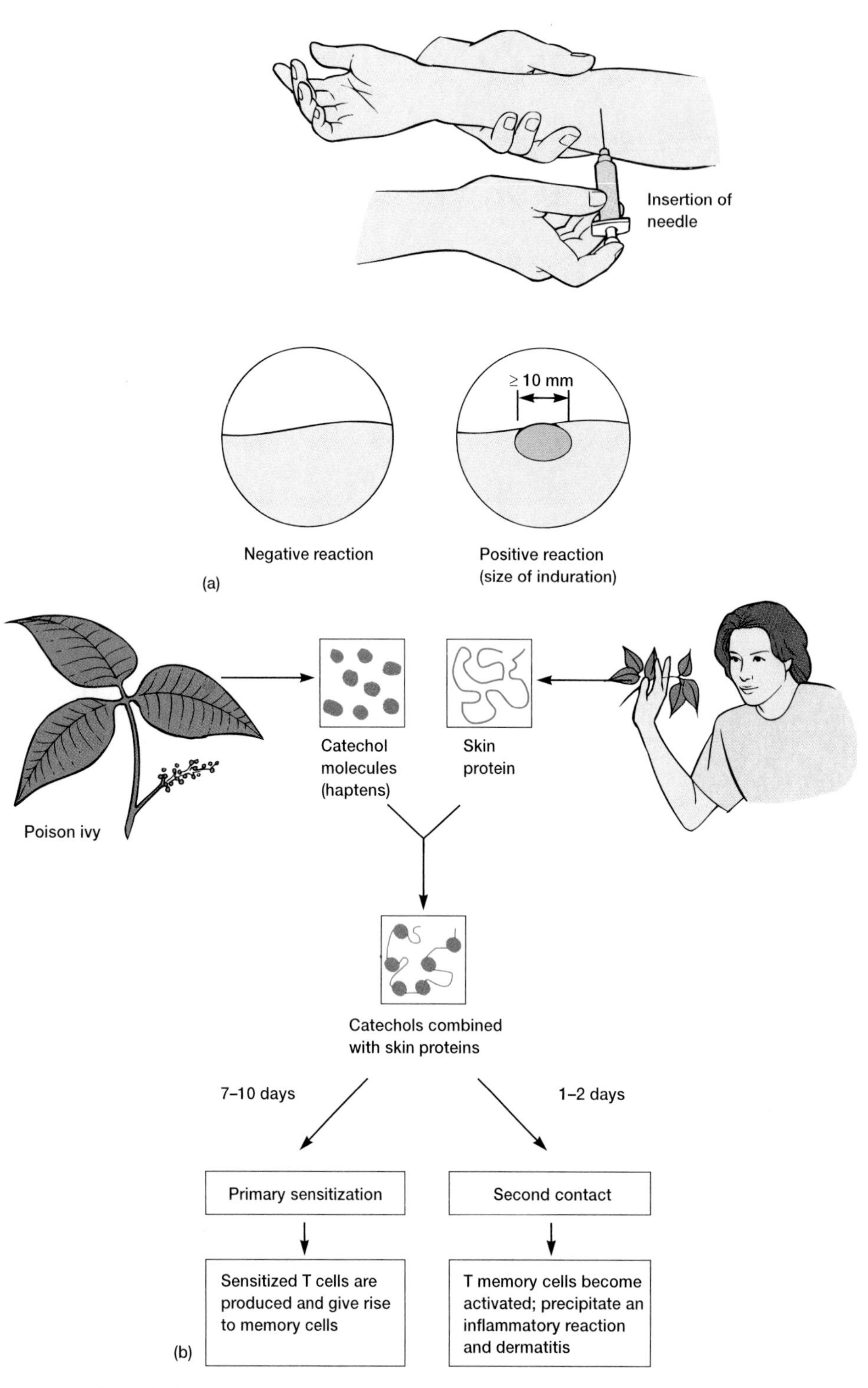

Insertion of
needle

≥ 10 mm

Negative reaction

Positive reaction
(size of induration)

(a)

Poison ivy

Catechol
molecules
(haptens)

Skin
protein

Catechols combined
with skin proteins

7–10 days

1–2 days

Primary sensitization

Second contact

Sensitized T cells are
produced and give rise
to memory cells

T memory cells become
activated; precipitate an
inflammatory reaction
and dermatitis

(b)

Type IV (Cell-Mediated) Hypersensitivity
Figure 31.14

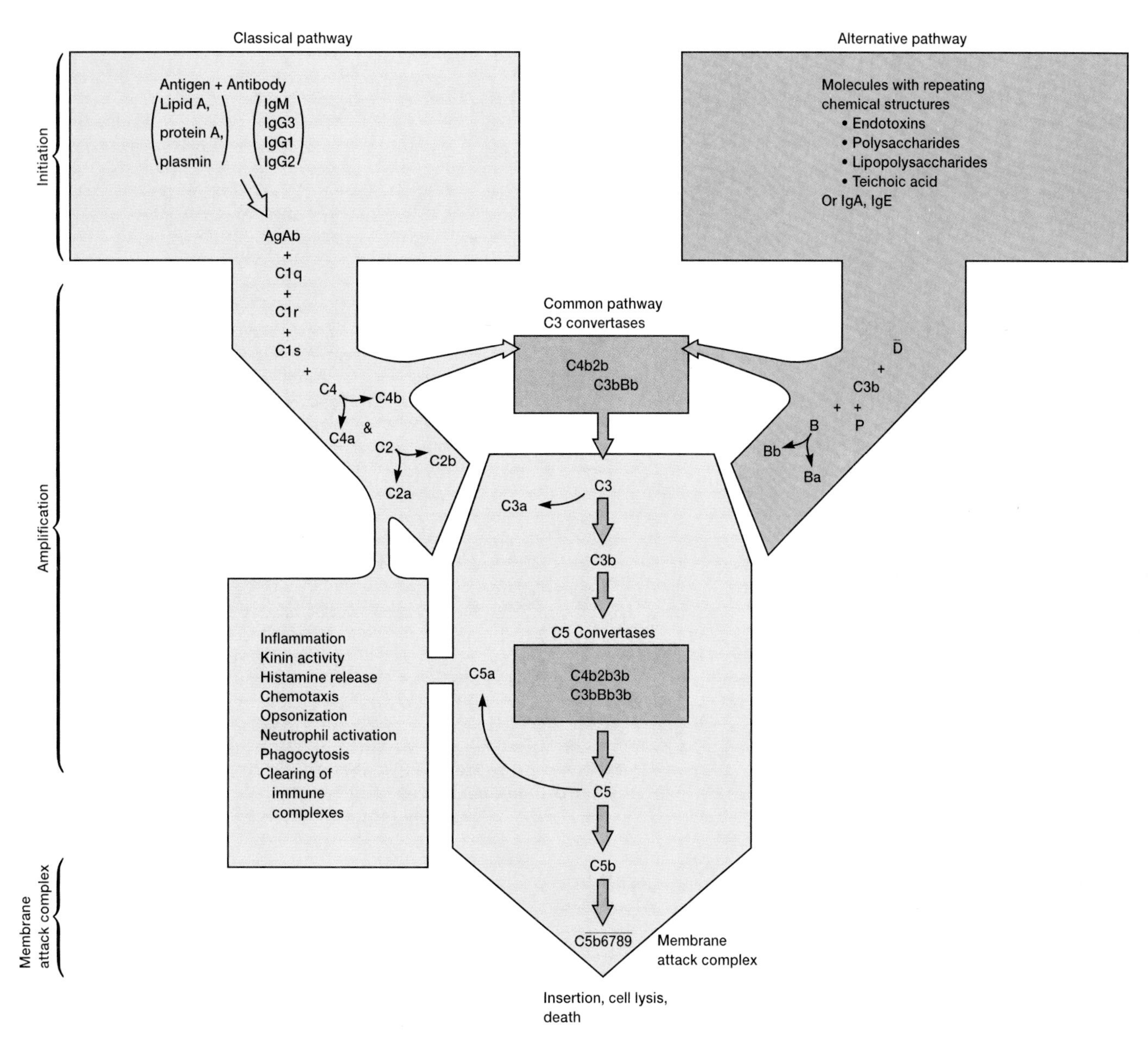

Classical pathway
Alternative pathway
Initiation
Amplification
Membrane attack complex
Antigen + Antibody
Lipid A,
protein A,
plasmin
IgM
IgG3
IgG1
IgG2
AgAb
+
C1q
+
C1r
+
C1s
+
C4
C4b
C4a
&
C2
C2b
C2a
Molecules with repeating chemical structures
• Endotoxins
• Polysaccharides
• Lipopolysaccharides
• Teichoic acid
Or IgA, IgE
D
+
C3b
+ +
B
P
Bb
Ba
Common pathway
C3 convertases
C4b2b
C3bBb
C3
C3a
C3b
C5 Convertases
C4b2b3b
C3bBb3b
C5a
C5
C5b
C5b6789
Membrane attack complex
Insertion, cell lysis, death
Inflammation
Kinin activity
Histamine release
Chemotaxis
Opsonization
Neutrophil activation
Phagocytosis
Clearing of immune complexes

Complement Cascade
Figure 32.2

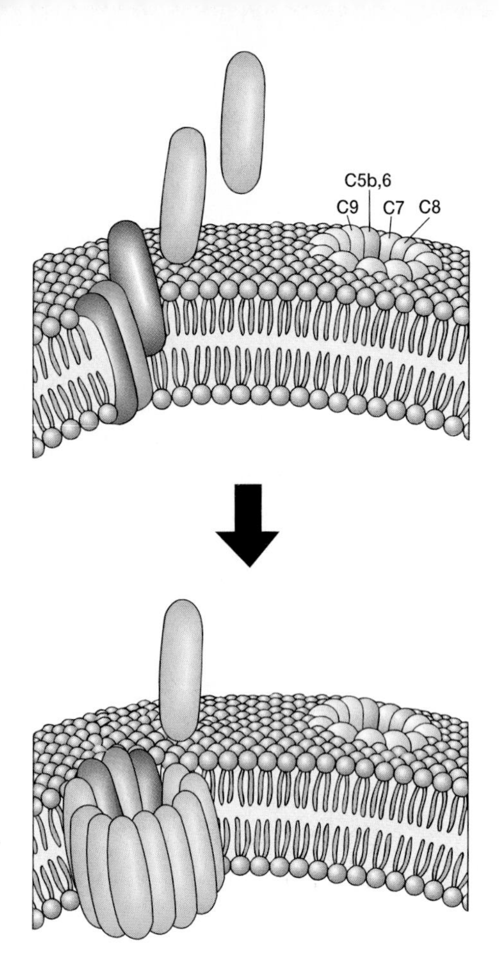

C5b,6
C9 | C7 | C8

Membrane Attack Complex
Figure 32.3

Phagocytic cell	Degree of binding	Opsonin
(a) Attachment by nonspecific receptors Microorganism	±	–
(b) Ab Fc receptor	+	Antibody
(c) C3b C3b receptor	+ +	Complement C3b
(d)	+ + + +	Antibody and complement C3b

Opsonization
Figure 32.6

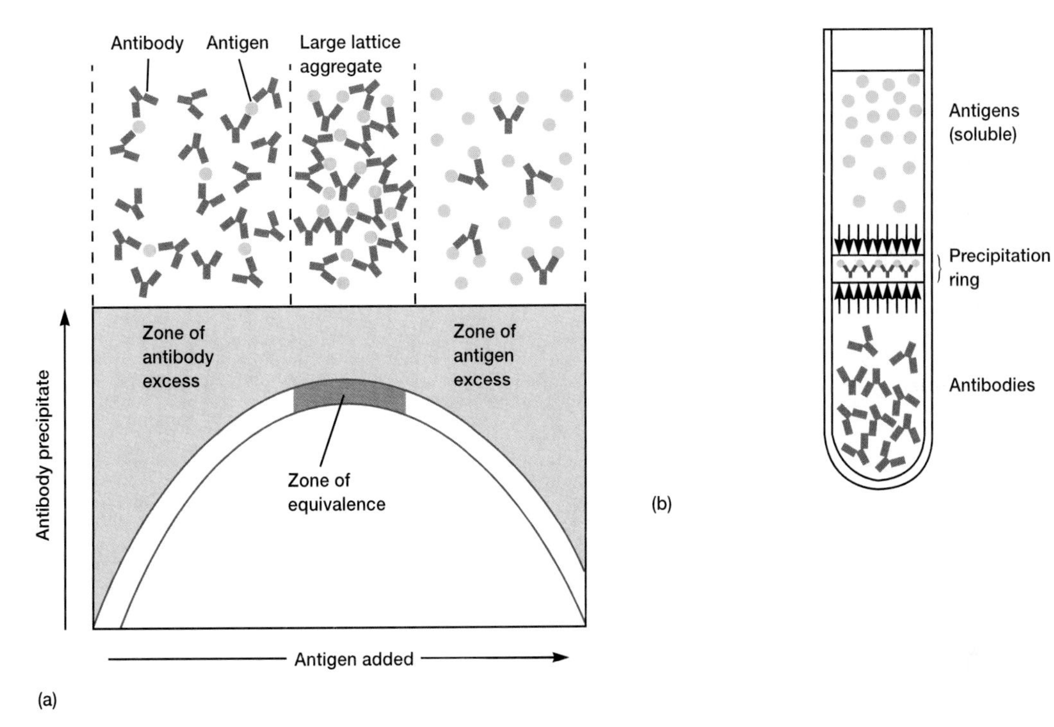
Antibody
Antigen
Large lattice aggregate
Zone of antibody excess
Zone of antigen excess
Zone of equivalence
Antibody precipitate
Antigen added
(a)
(b)
Antigens (soluble)
Precipitation ring
Antibodies

Immunoprecipitation
Figure 32.17

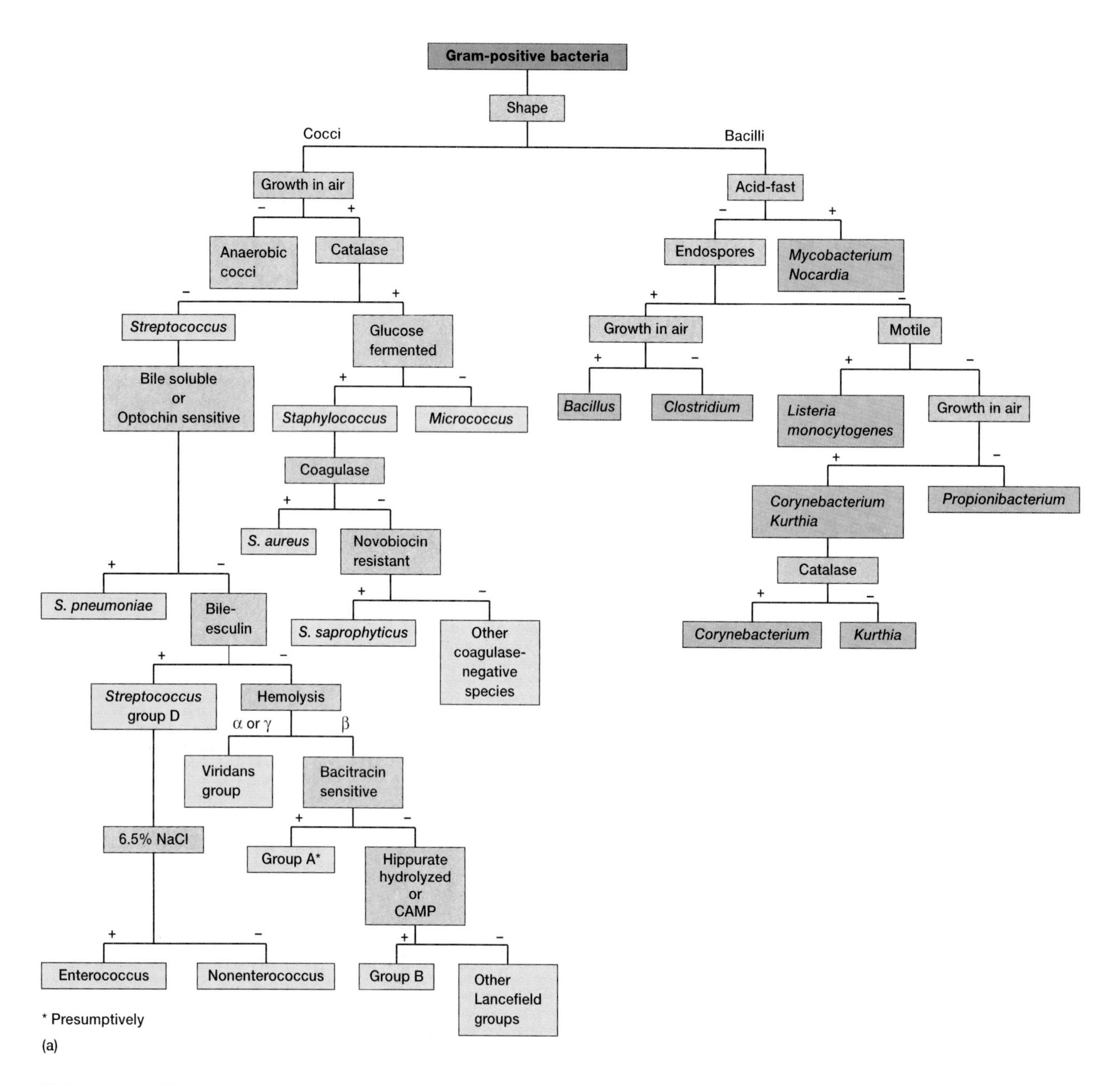

Dichotomous Keys for Clinically Important Genera
Figure 34.7 *a*

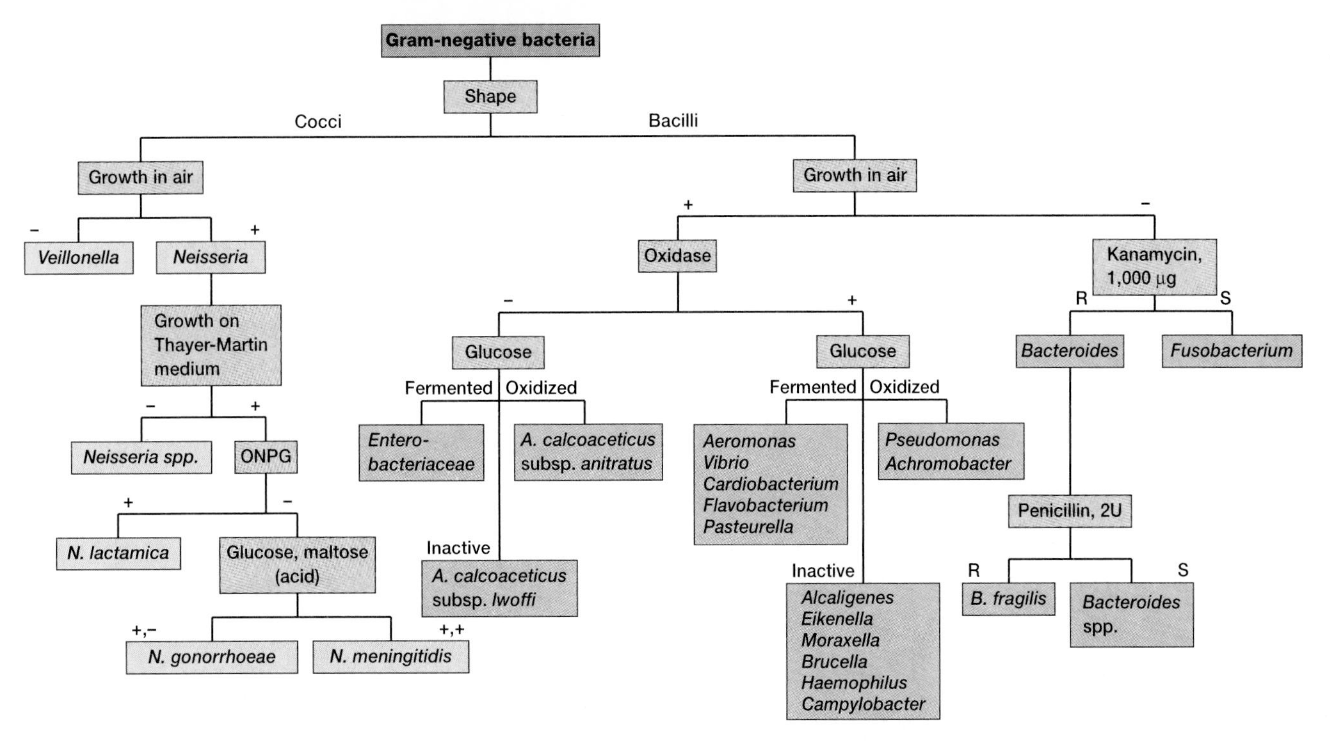

Gram-negative bacteria

Dichotomous Keys for Clinically Important Genera
Figure 34.7 *b*

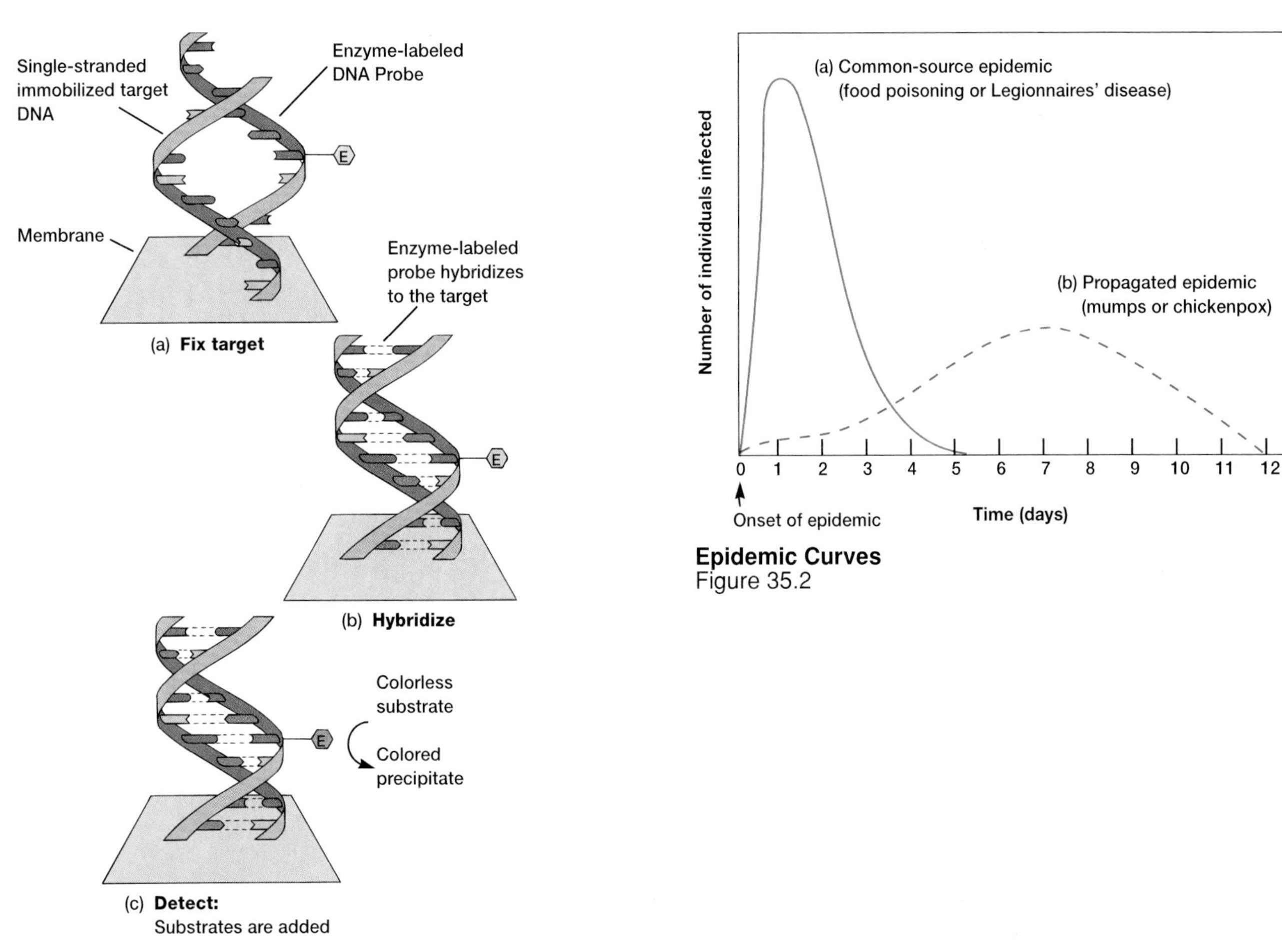

Single-stranded immobilized target DNA

Enzyme-labeled DNA Probe

Membrane

(a) **Fix target**

Enzyme-labeled probe hybridizes to the target

(b) **Hybridize**

Colorless substrate

Colored precipitate

(c) **Detect:**
Substrates are added

DNA Probe Hybridization
Figure 34.13

(a) Common-source epidemic (food poisoning or Legionnaires' disease)

(b) Propagated epidemic (mumps or chickenpox)

Number of individuals infected

Onset of epidemic

Time (days)

Epidemic Curves
Figure 35.2

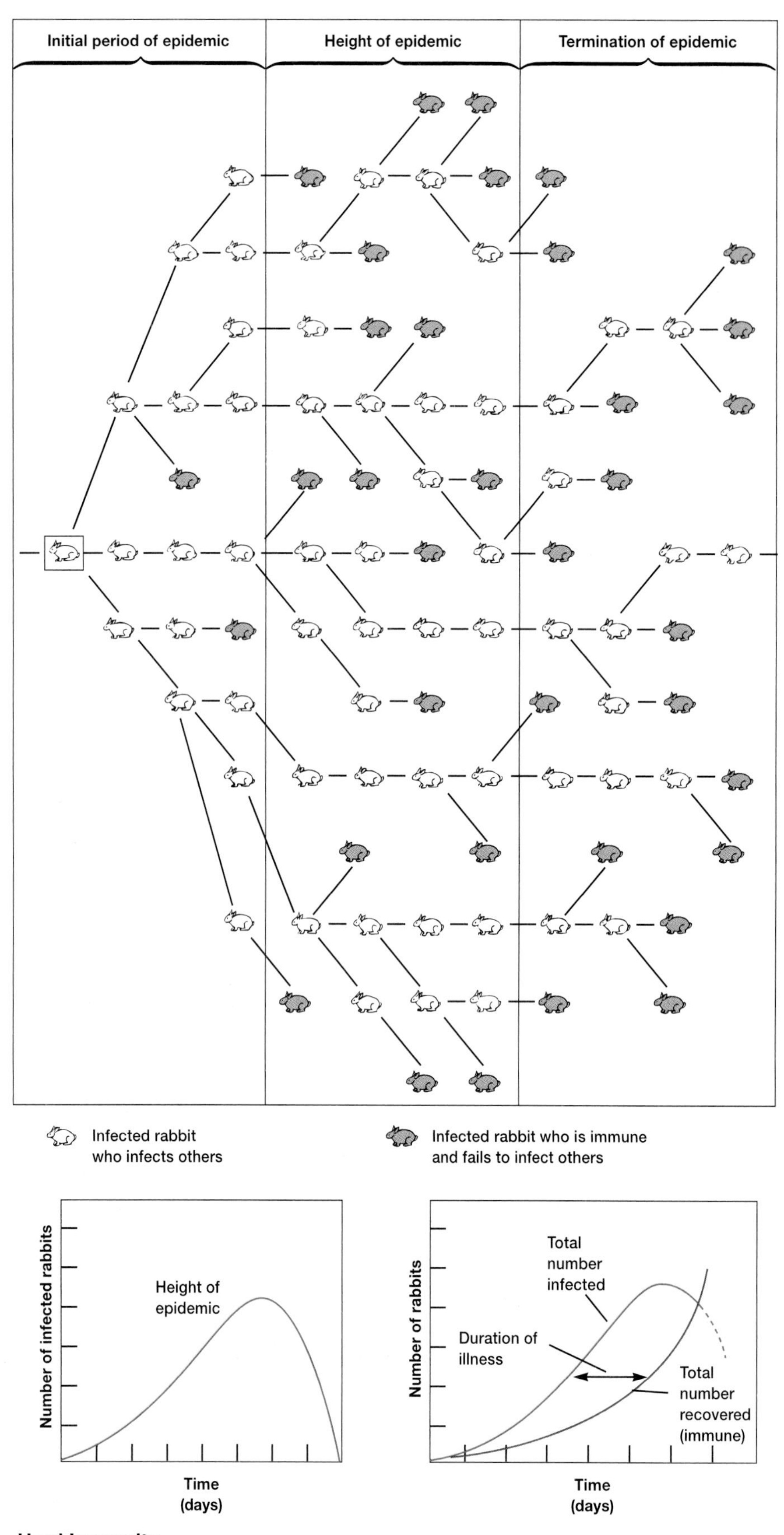

Initial period of epidemic	Height of epidemic	Termination of epidemic

Infected rabbit who infects others

Infected rabbit who is immune and fails to infect others

Number of infected rabbits

Height of epidemic

Time (days)

Number of rabbits

Total number infected

Duration of illness

Total number recovered (immune)

Time (days)

Herd Immunity
Figure 35.4

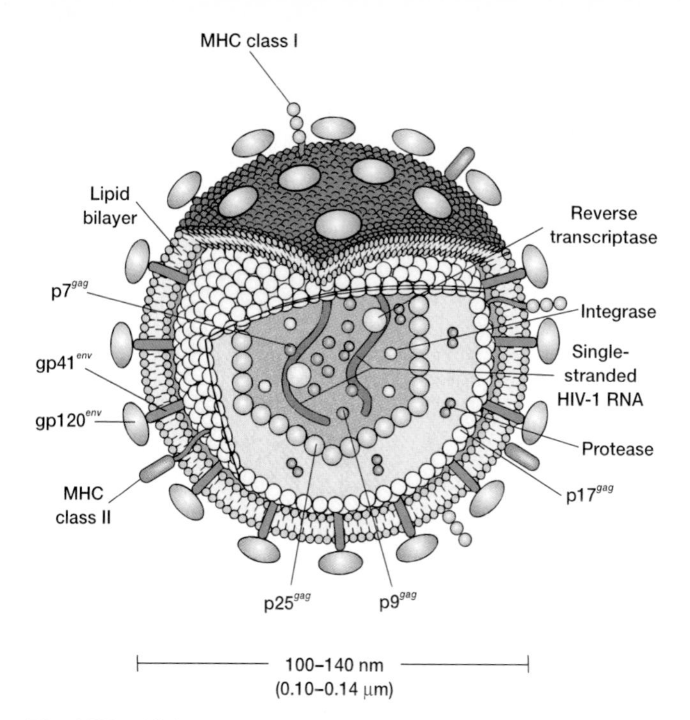

MHC class I

Lipid bilayer

p7^{gag}

gp41^{env}

gp120^{env}

MHC class II

Reverse transcriptase

Integrase

Single-stranded HIV-1 RNA

Protease

p17^{gag}

p25^{gag} p9^{gag}

100–140 nm
(0.10–0.14 µm)

The HIV-1 Virion
Figure 36.8

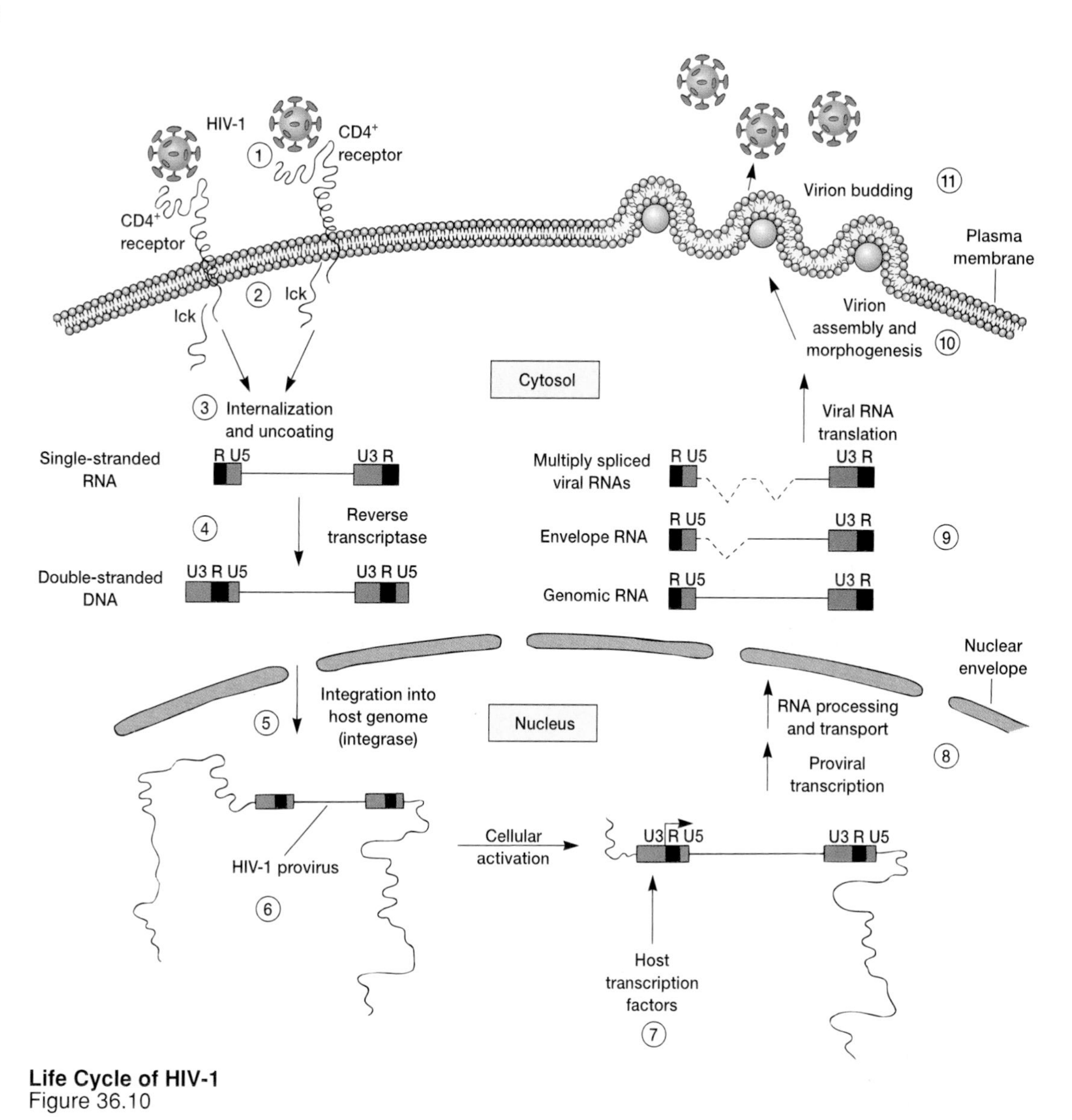

Life Cycle of HIV-1
Figure 36.10

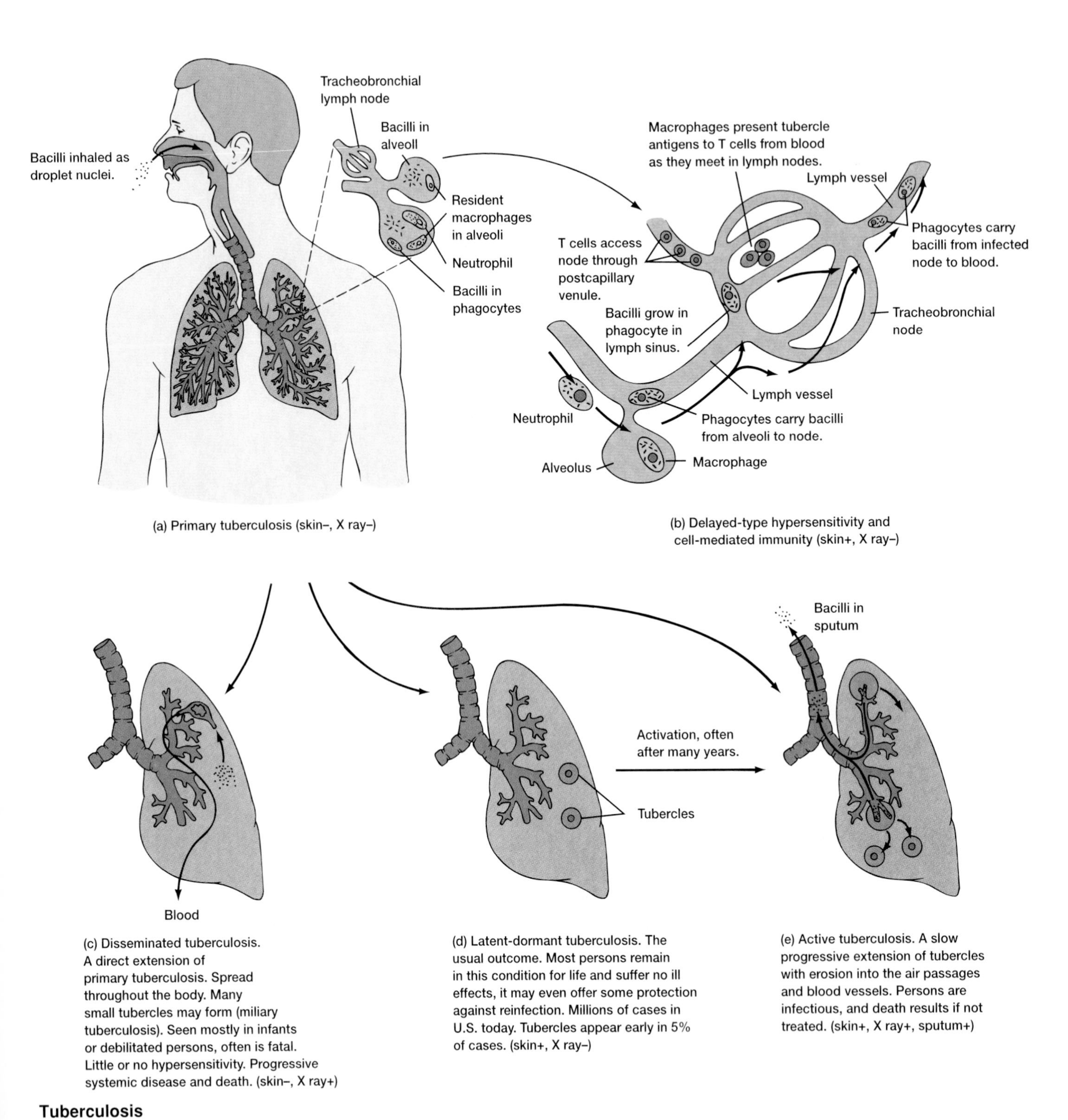

Tracheobronchial lymph node

Bacilli in alveoll

Bacilli inhaled as droplet nuclei.

Resident macrophages in alveoli

Neutrophil

Bacilli in phagocytes

Macrophages present tubercle antigens to T cells from blood as they meet in lymph nodes.

Lymph vessel

T cells access node through postcapillary venule.

Phagocytes carry bacilli from infected node to blood.

Bacilli grow in phagocyte in lymph sinus.

Tracheobronchial node

Lymph vessel

Neutrophil

Phagocytes carry bacilli from alveoli to node.

Alveolus

Macrophage

(a) Primary tuberculosis (skin–, X ray–)

(b) Delayed-type hypersensitivity and cell-mediated immunity (skin+, X ray–)

Blood

Activation, often after many years.

Tubercles

Bacilli in sputum

(c) Disseminated tuberculosis. A direct extension of primary tuberculosis. Spread throughout the body. Many small tubercles may form (miliary tuberculosis). Seen mostly in infants or debilitated persons, often is fatal. Little or no hypersensitivity. Progressive systemic disease and death. (skin–, X ray+)

(d) Latent-dormant tuberculosis. The usual outcome. Most persons remain in this condition for life and suffer no ill effects, it may even offer some protection against reinfection. Millions of cases in U.S. today. Tubercles appear early in 5% of cases. (skin+, X ray–)

(e) Active tuberculosis. A slow progressive extension of tubercles with erosion into the air passages and blood vessels. Persons are infectious, and death results if not treated. (skin+, X ray+, sputum+)

Tuberculosis
Figure 37.7

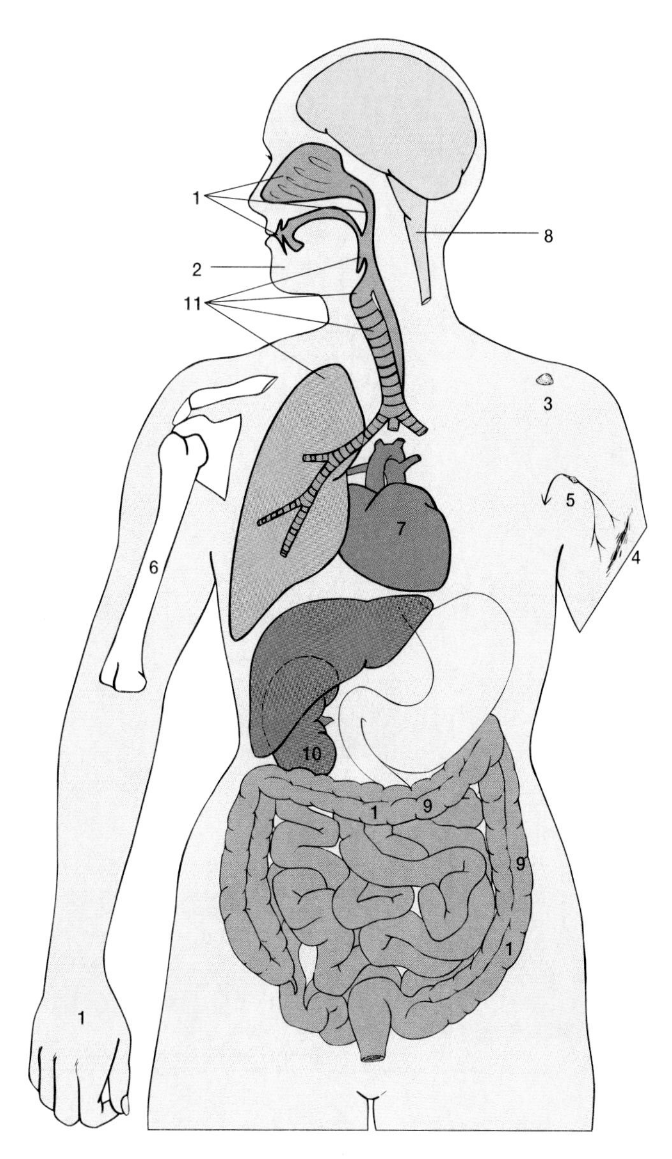

1 Tissue where *S. aureus*
  is often found but does not
  normally cause disease

Diseases that may be
caused by *S. aureus* are:

2 Pimples and impetigo

3 Boils and carbuncles
  on any surface area

4 Wound infections and
  abscesses

5 Spread to lymph nodes
  and to blood (septicemia),
  resulting in widespread
  seeding

6 Osteomyelitis

7 Endocarditis

8 Meningitis

9 Enteritis and enterotoxin
  poisoning (food poisoning)

10 Nephritis

11 Respiratory infections:
   Pharyngitis
   Laryngitis
   Bronchitis
   Pneumonia

Staphylococcal Diseases
Figure 37.16

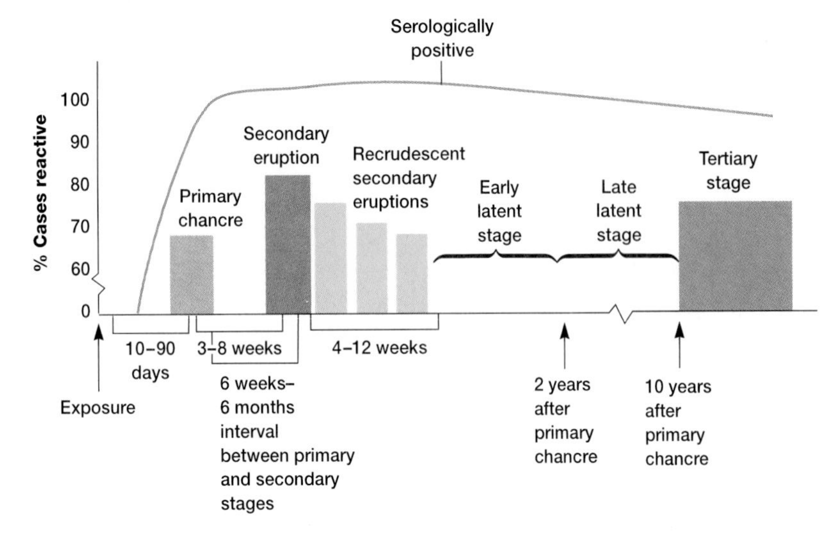

The Course of Untreated Syphilis
Figure 37.19

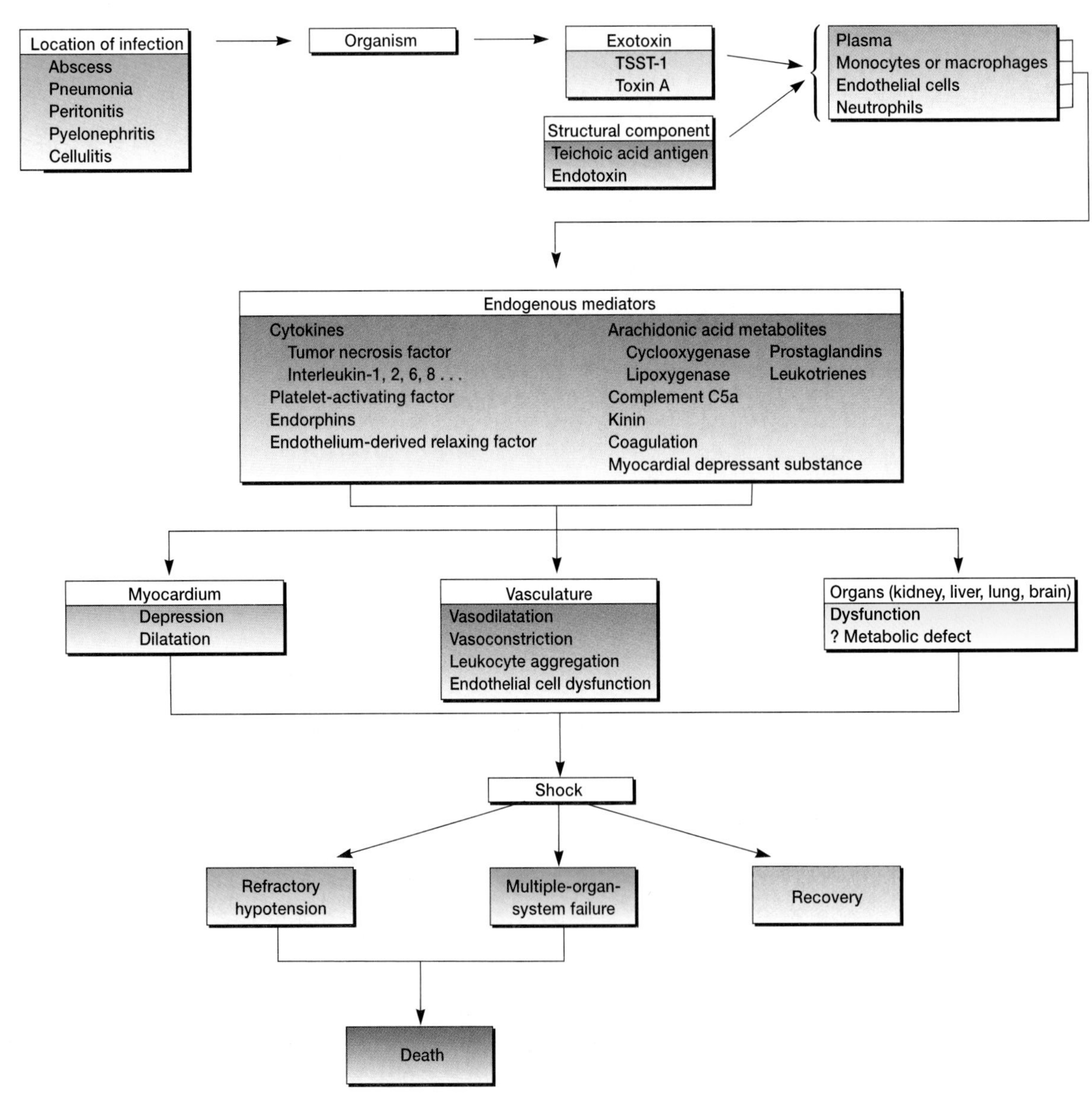

The Septic Shock Cascade
Figure 37.22

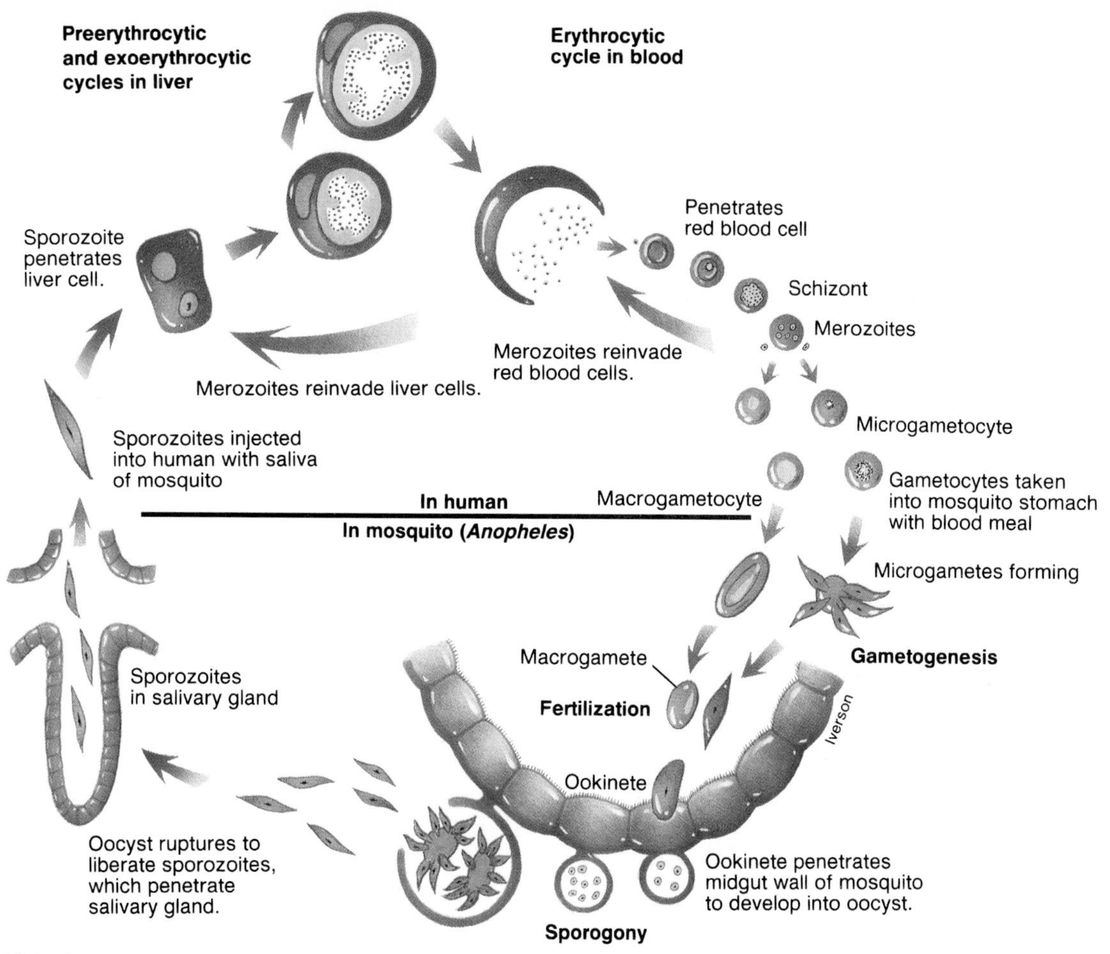

Preerythrocytic and exoerythrocytic cycles in liver

Erythrocytic cycle in blood

Sporozoite penetrates liver cell.

Penetrates red blood cell

Schizont

Merozoites

Merozoites reinvade red blood cells.

Merozoites reinvade liver cells.

Sporozoites injected into human with saliva of mosquito

Microgametocyte

Gametocytes taken into mosquito stomach with blood meal

In human

Macrogametocyte

In mosquito (Anopheles)

Microgametes forming

Sporozoites in salivary gland

Macrogamete

Fertilization

Gametogenesis

Iverson

Ookinete

Oocyst ruptures to liberate sporozoites, which penetrate salivary gland.

Ookinete penetrates midgut wall of mosquito to develop into oocyst.

Sporogony

Malaria
Figure 39.19

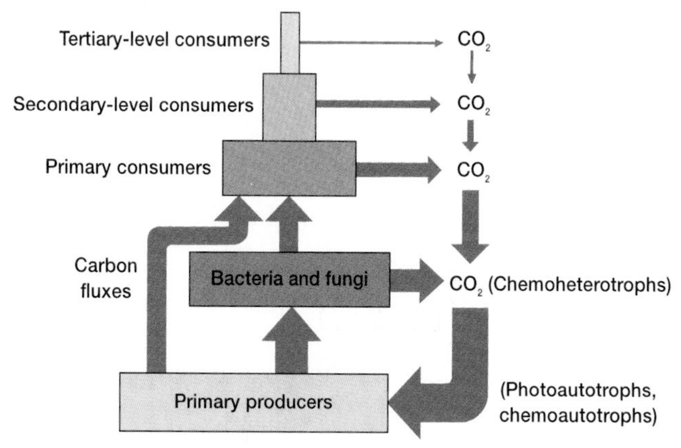

Tertiary-level consumers → CO_2

Secondary-level consumers → CO_2

Primary consumers → CO_2

Carbon fluxes

Bacteria and fungi → CO_2 (Chemoheterotrophs)

Primary producers

(Photoautotrophs, chemoautotrophs)

Ecological Role of Microorganisms
Figure 40.2

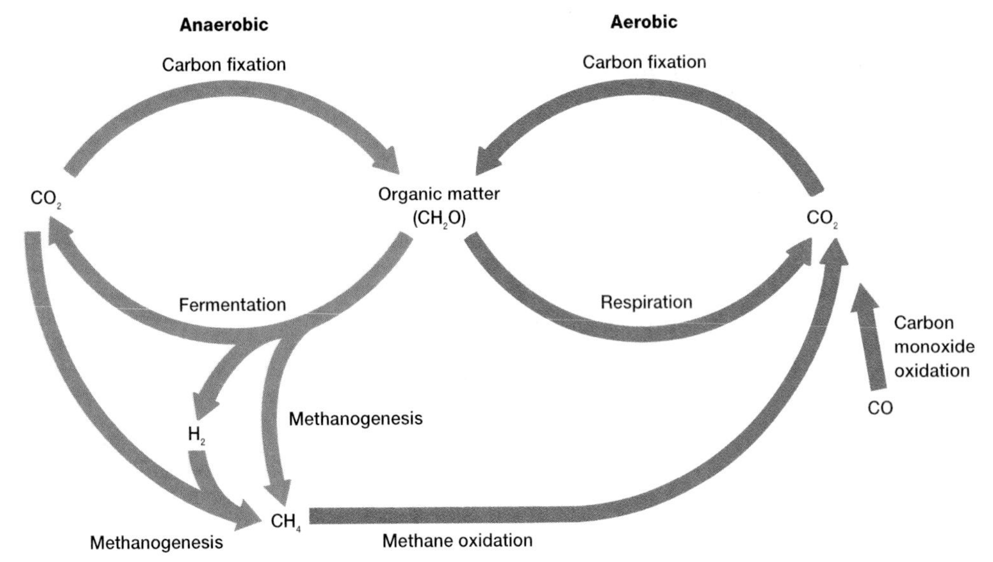

The Carbon Cycle in the Environment
Figure 40.7

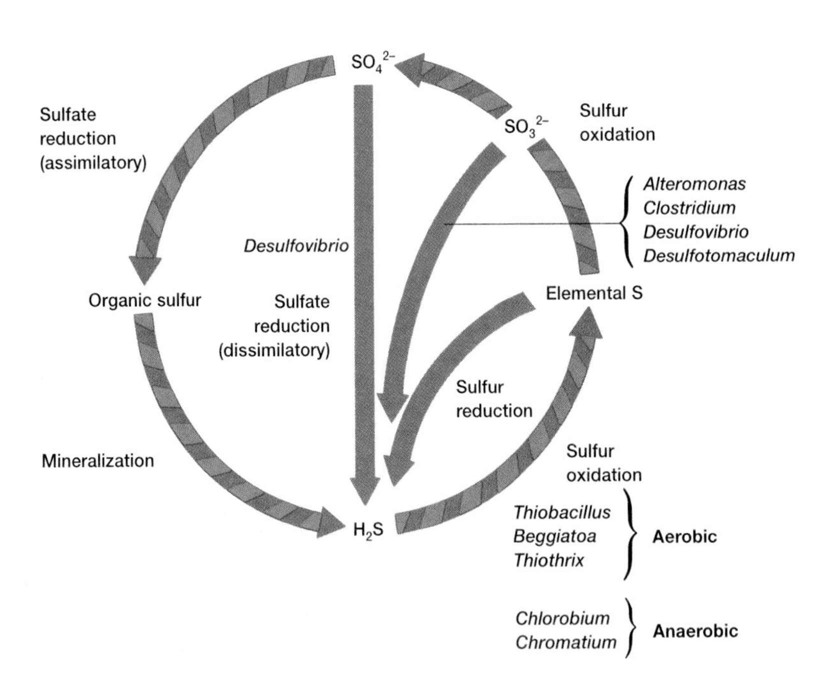

The Sulfur Cycle
Figure 40.8

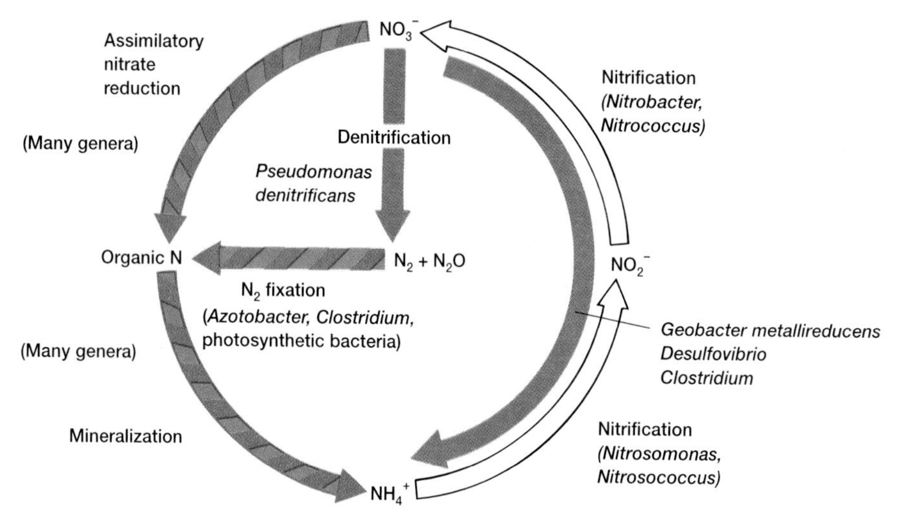

The Environmental Nitrogen Cycle
Figure 40.9

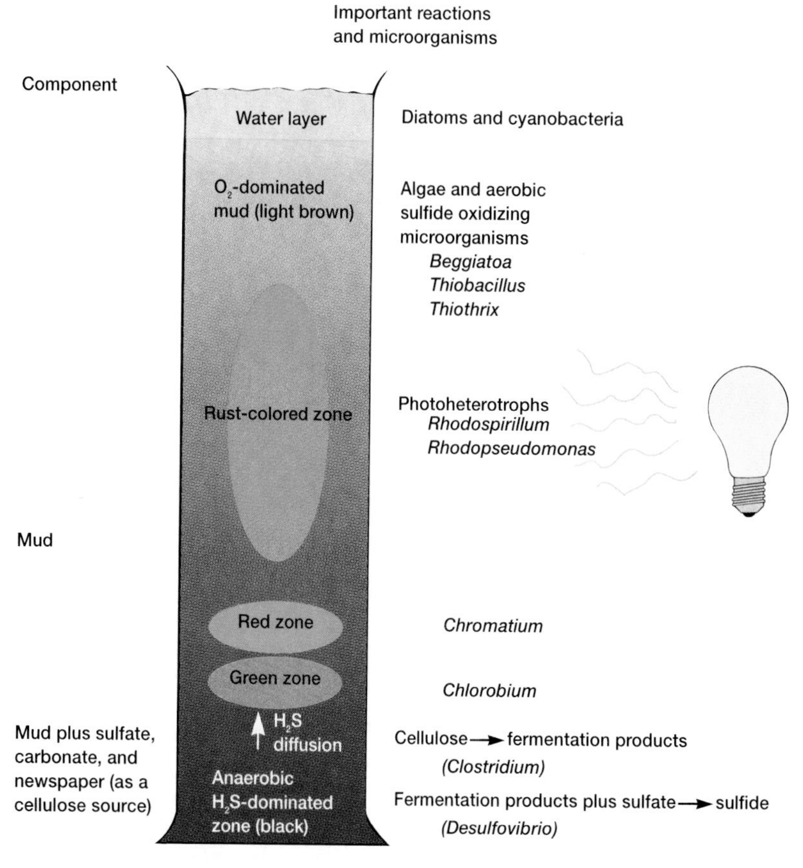

The Winogradsky Column
Figure 41.1

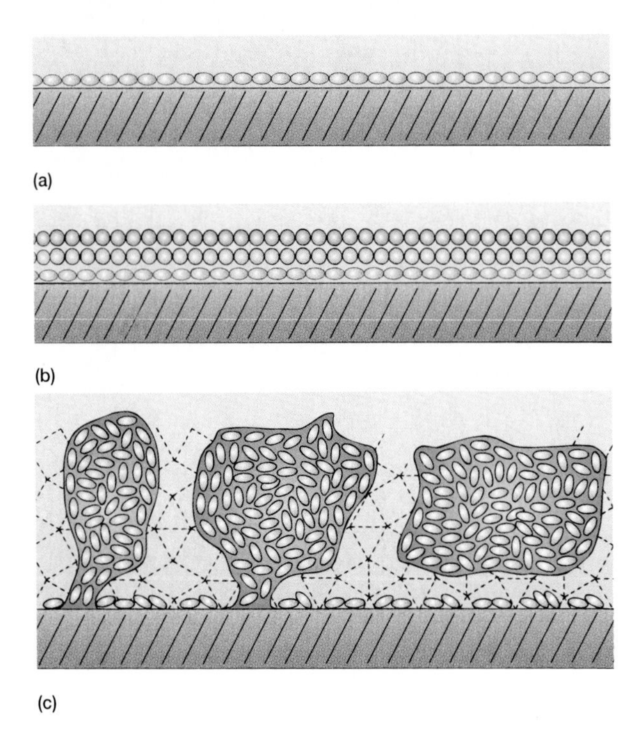

(a)

(b)

(c)

The Growth of Biofilms
Figure 41.3

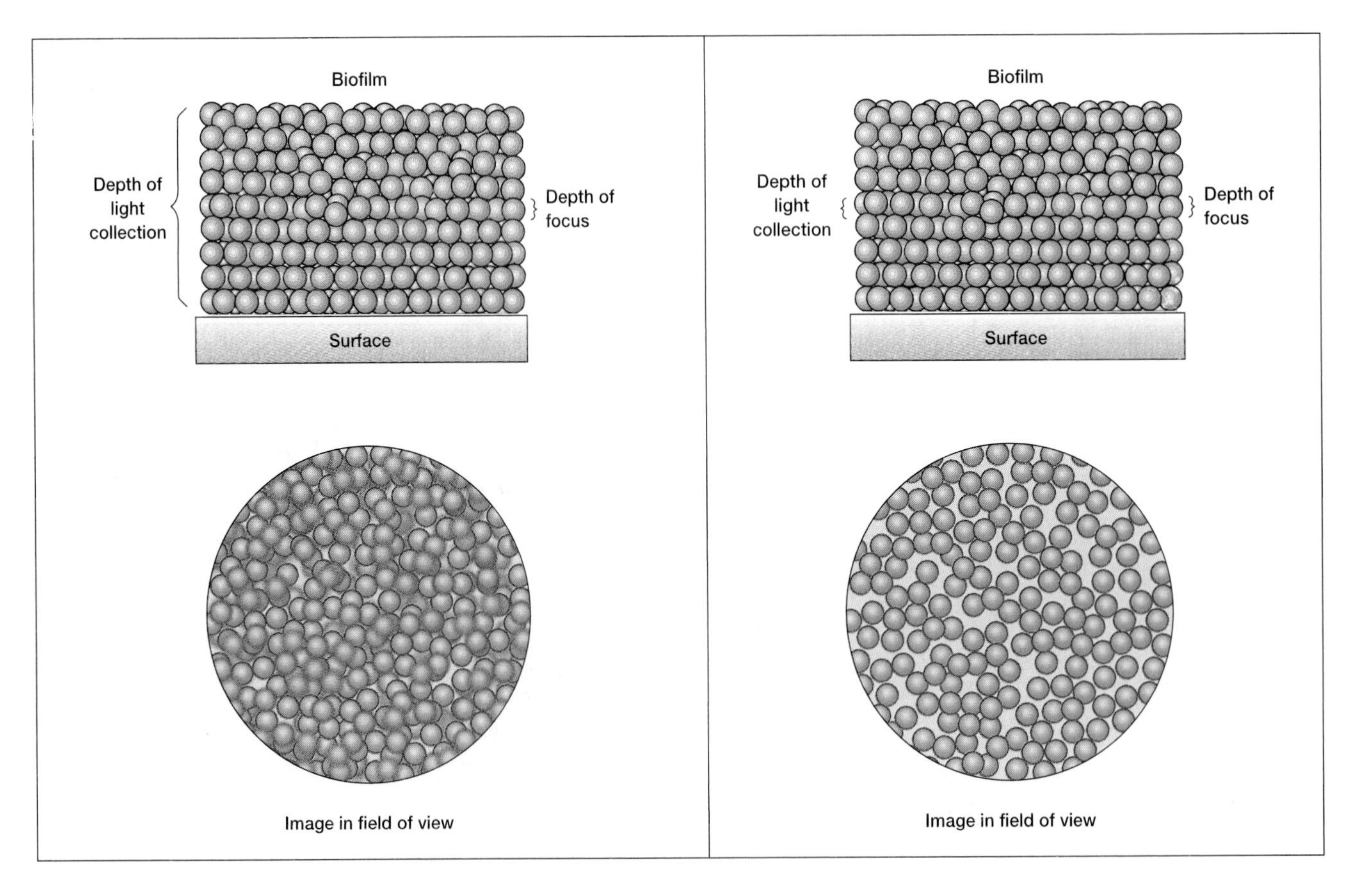

Confocal Scanning Laser Microscopy
Box Figure 41.1

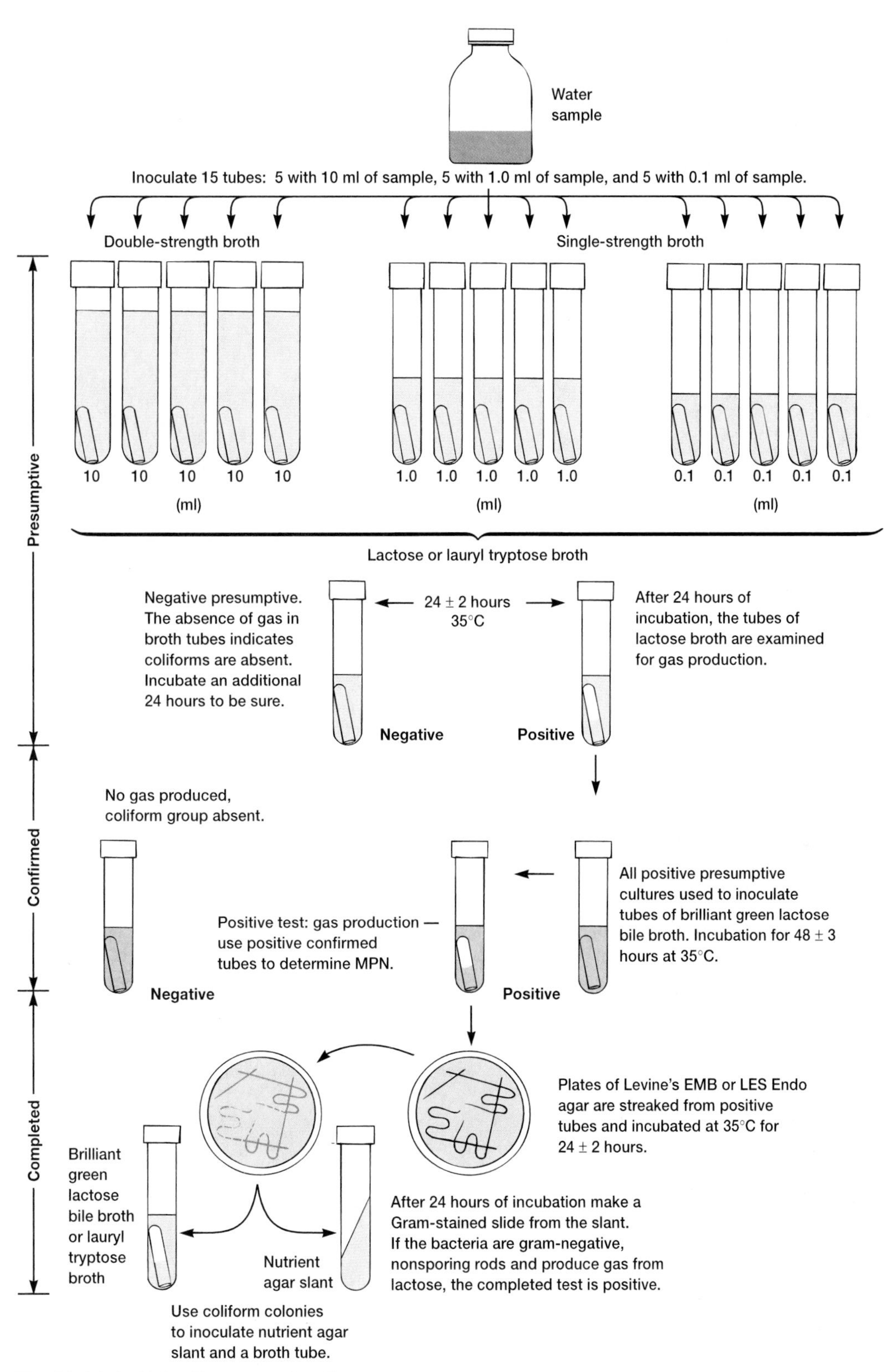

Water
sample

Inoculate 15 tubes: 5 with 10 ml of sample, 5 with 1.0 ml of sample, and 5 with 0.1 ml of sample.

Double-strength broth

Single-strength broth

10 10 10 10 10
(ml)

1.0 1.0 1.0 1.0 1.0
(ml)

0.1 0.1 0.1 0.1 0.1
(ml)

Presumptive

Lactose or lauryl tryptose broth

Negative presumptive.
The absence of gas in
broth tubes indicates
coliforms are absent.
Incubate an additional
24 hours to be sure.

24 ± 2 hours
35°C

After 24 hours of
incubation, the tubes of
lactose broth are examined
for gas production.

Negative

Positive

Confirmed

No gas produced,
coliform group absent.

Positive test: gas production —
use positive confirmed
tubes to determine MPN.

All positive presumptive
cultures used to inoculate
tubes of brilliant green lactose
bile broth. Incubation for 48 ± 3
hours at 35°C.

Negative

Positive

Completed

Brilliant
green
lactose
bile broth
or lauryl
tryptose
broth

Nutrient
agar slant

Plates of Levine's EMB or LES Endo
agar are streaked from positive
tubes and incubated at 35°C for
24 ± 2 hours.

After 24 hours of incubation make a
Gram-stained slide from the slant.
If the bacteria are gram-negative,
nonsporing rods and produce gas from
lactose, the completed test is positive.

Use coliform colonies
to inoculate nutrient agar
slant and a broth tube.

The Multiple-Tube Fermentation Test
Figure 41.21

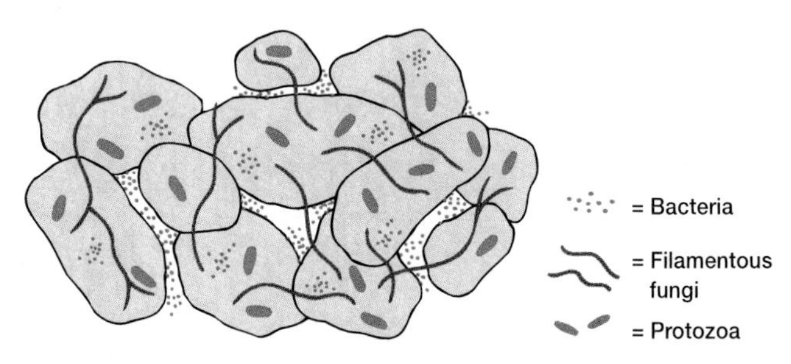

The Microenvironment—The World of Microorganisms in Soil
Figure 42.1

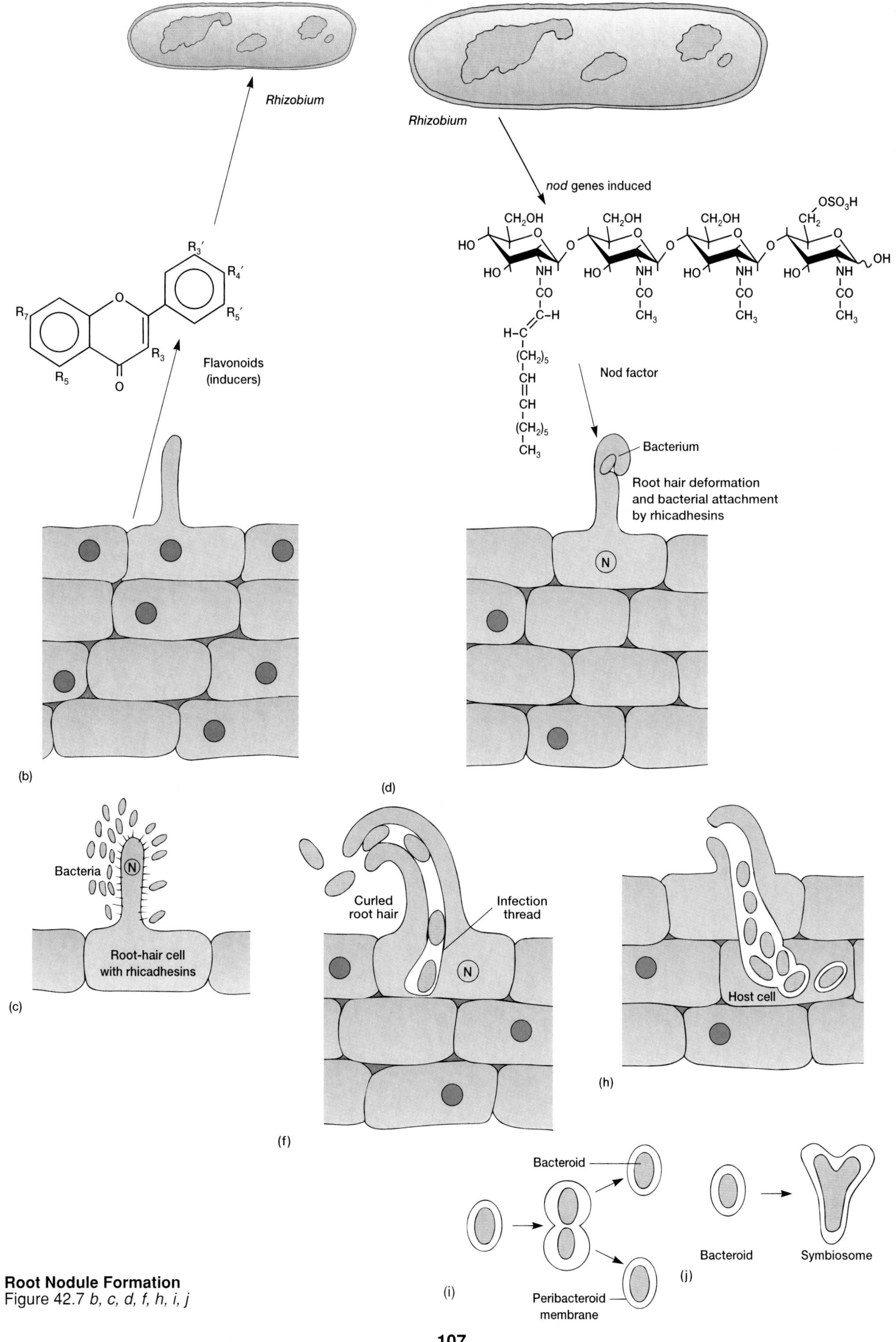

Rhizobium

Rhizobium

nod genes induced

R_3'
R_4'
R_7
R_3
R_5'
R_5

Flavonoids
(inducers)

Nod factor

Bacterium

Root hair deformation
and bacterial attachment
by rhicadhesins

(N)

(b)

(d)

Bacteria
(N)

Root-hair cell
with rhicadhesins

(c)

Curled
root hair

Infection
thread

(N)

(f)

Host cell

(h)

Bacteroid

Peribacteroid
membrane

(i)

Bacteroid

Symbiosome

(j)

Root Nodule Formation
Figure 42.7 *b, c, d, f, h, i, j*

Processing step   Biological change

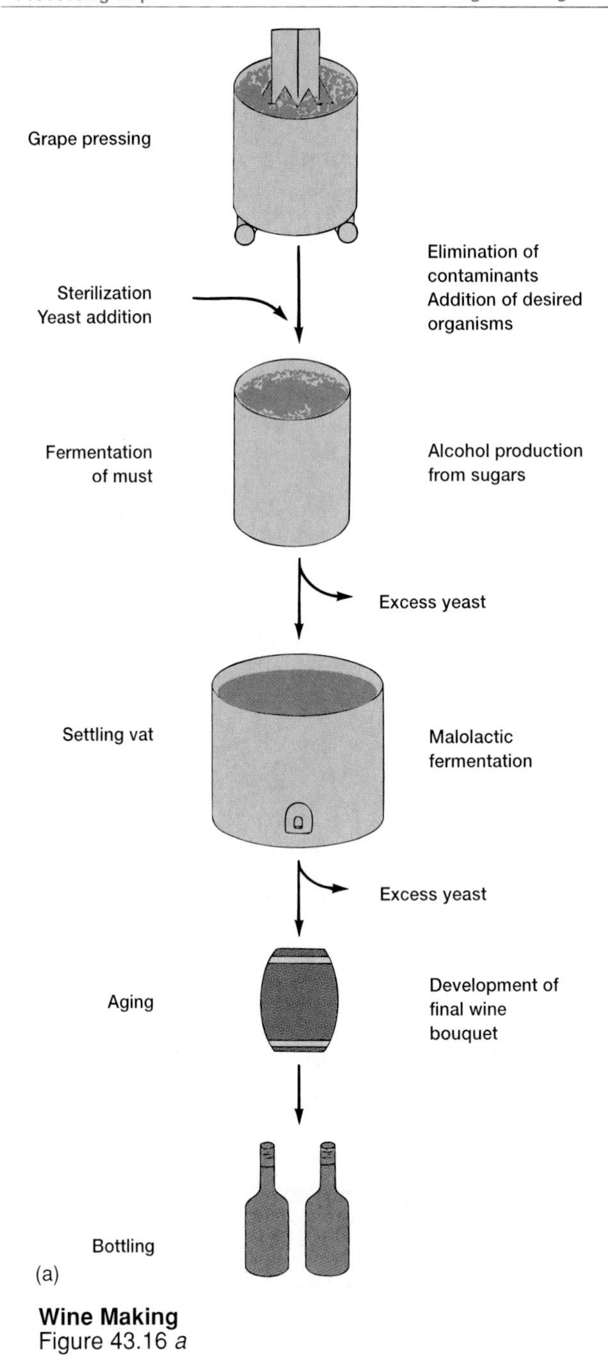

Grape pressing

Sterilization   Elimination of
Yeast addition   contaminants
                       Addition of desired
                       organisms

Fermentation   Alcohol production
of must   from sugars

                       Excess yeast

Settling vat   Malolactic
                       fermentation

                       Excess yeast

Aging   Development of
                       final wine
                       bouquet

Bottling
(a)

Wine Making
Figure 43.16 *a*

Processing step   Biological change

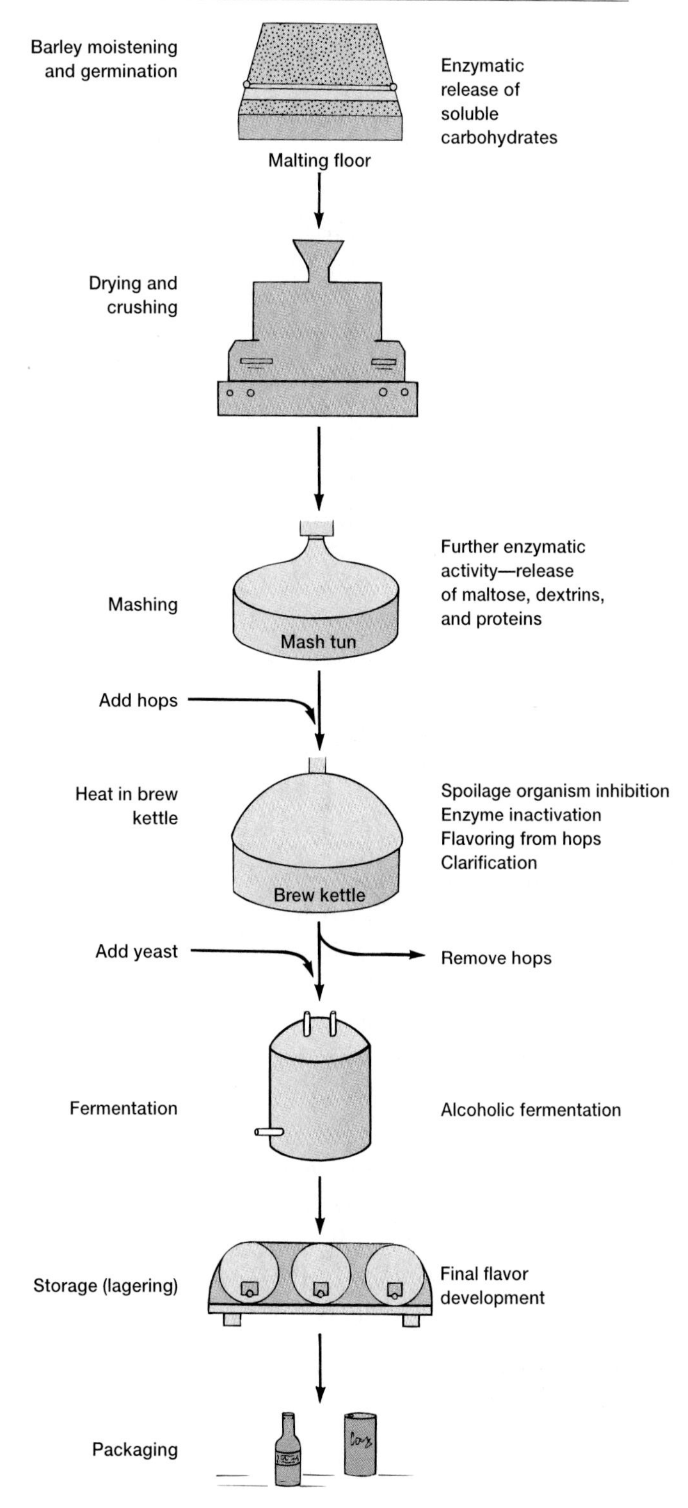

Barley moistening
and germination   Enzymatic
                                                                   release of
                                                                   soluble
                                                                   carbohydrates

Malting floor

Drying and
crushing

Mashing   Further enzymatic
                                                                   activity—release
                                                                   of maltose, dextrins,
                                                                   and proteins
Mash tun

Add hops

Heat in brew   Spoilage organism inhibition
kettle   Enzyme inactivation
                                                                   Flavoring from hops
                                                                   Clarification
Brew kettle

Add yeast   Remove hops

Fermentation   Alcoholic fermentation

Storage (lagering)   Final flavor
                                                                   development

Packaging

Producing Beer
Figure 43.17

109

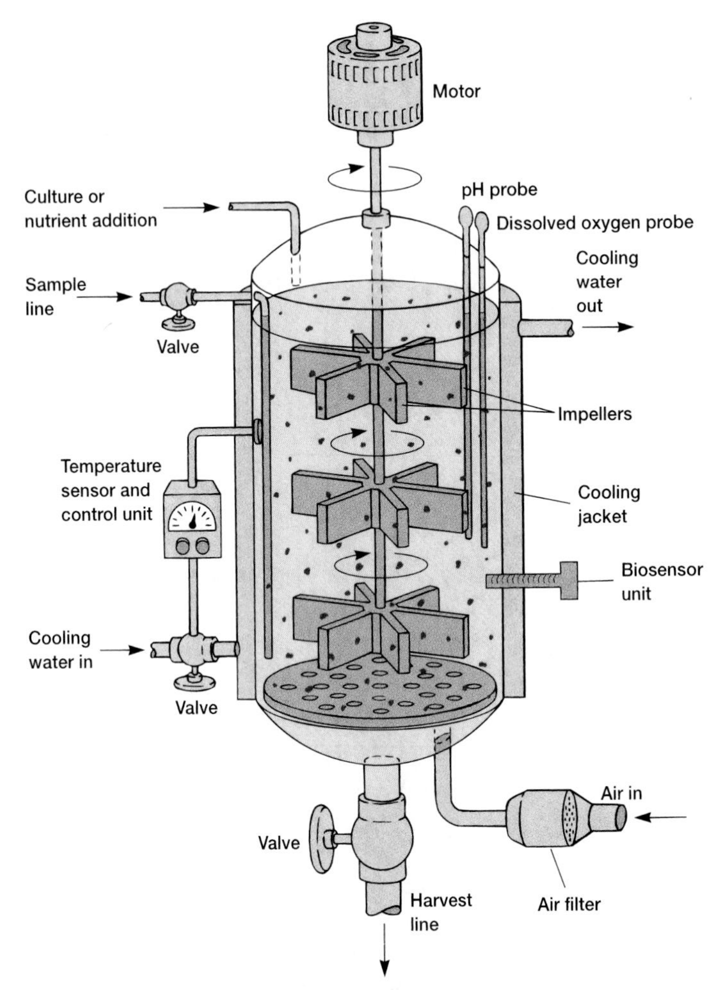

Motor

Culture or
nutrient addition

pH probe

Dissolved oxygen probe

Cooling
water
out

Sample
line

Valve

Impellers

Temperature
sensor and
control unit

Cooling
jacket

Biosensor
unit

Cooling
water in

Valve

Valve

Harvest
line

Air in

Air filter

A Large-Scale Fermentation Unit
Figure 44.2

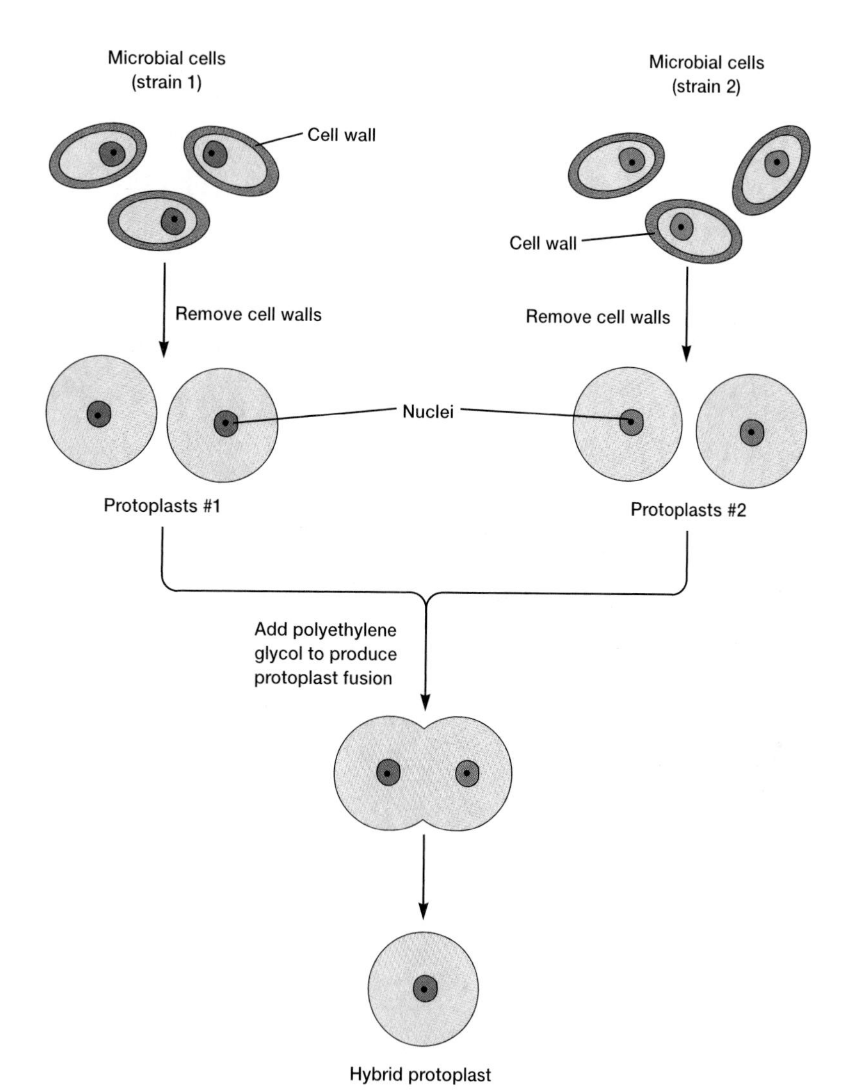

Microbial cells
(strain 1)

Cell wall

Microbial cells
(strain 2)

Cell wall

Remove cell walls

Remove cell walls

Nuclei

Protoplasts #1

Protoplasts #2

Add polyethylene
glycol to produce
protoplast fusion

Hybrid protoplast

**Protoplast Fusion and the Genetic Manipulation of
Microorganisms**
Figure 44.6

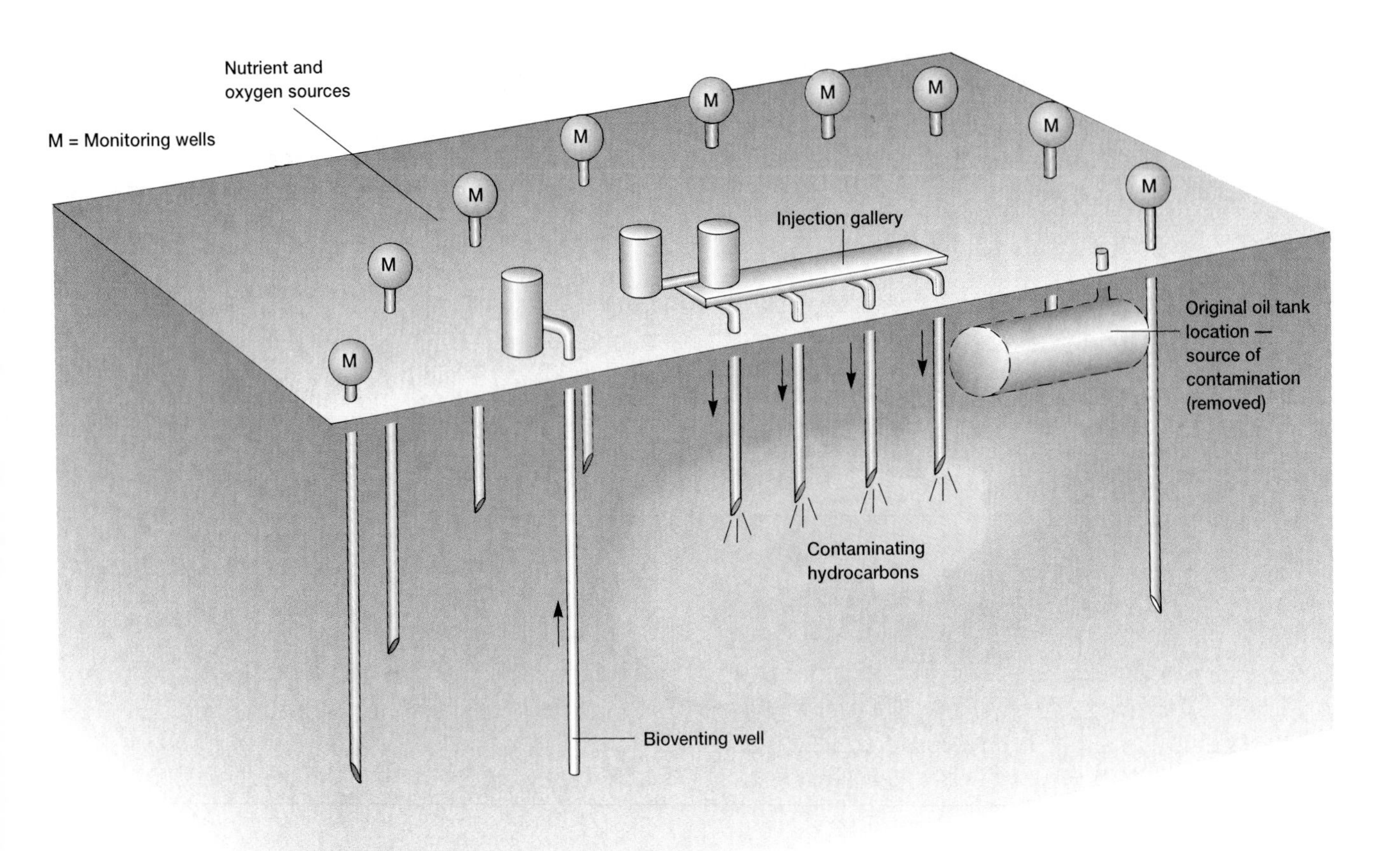

Nutrient and
oxygen sources
M = Monitoring wells
M
Injection gallery
Original oil tank
location —
source of
contamination
(removed)
Contaminating
hydrocarbons
Bioventing well

An in situ Bioremediation System
Figure 44.21

Line Art

Fig. 10.19 Art: From Geoffrey Zubay, *Biochemistry,* 2d edition. Copyright © 1988 Macmillan Publishing Co. Reprinted by permission of Times Mirror Higher Education Group, Inc., Dubuque, Iowa. All Rights Reserved.

Fig. 12.18 Art: From Benjamin Lewin, *Genes,* 4th edition. Copyright © 1990 Cell Press, Cambridge, MA. Reprinted by permission.

Fig. 14.13 Art: From Leland G. Johnson, *Biology,* 2d edition. Copyright © 1987 Wm. C. Brown Communications, Inc. Reprinted by permission of Times Mirror Higher Education Group, Inc., Dubuque, Iowa. All Rights Reserved.

Fig. 14.14 Art: From Leland G. Johnson, *Biology,* 2d edition. Copyright © 1987 Wm. C. Brown Communications, Inc. Reprinted by permission of Times Mirror Higher Education Group, Inc., Dubuque, Iowa. All Rights Reserved.

Fig. 14.15 Art: From Leland G. Johnson, *Biology,* 2d edition. Copyright © 1987 Wm. C. Brown Communications, Inc. Reprinted by permission of Times Mirror Higher Education Group, Inc., Dubuque, Iowa. All Rights Reserved.

Fig. 14.17 Art: From Robert F. Weaver and Philip W. Hedrick, *Genetics,* 2d edition. Copyright © 1992 Wm. C. Brown Communications, Inc. Reprinted by permission of Times Mirror Higher Education Group, Inc., Dubuque, Iowa. All Rights Reserved.

Fig. 14.18 Art: From Leland G. Johnson, *Biology,* 2d edition. Copyright © 1987 Wm. C. Brown Communications, Inc. Reprinted by permission of Times Mirror Higher Education Group, Inc., Dubuque, Iowa. All Rights Reserved.

Fig. 14.19 Art: From Leland G. Johnson, *Biology,* 2d edition. Copyright © 1987 Wm. C. Brown Communications, Inc. Reprinted by permission of Times Mirror Higher Education Group, Inc., Dubuque, Iowa. All Rights Reserved.

Fig. 16.13 Art: *The Structure of an Icosahedral Capsid* from *Microbiology,* Third edition by Bernard D. Davis, et al. Copyright © 1980 by Harper & Row, Publishers, Inc. Reprinted by permission of HarperCollins Publishers, Inc.

Fig. 17.1 Art: From R. I. B. Francki, et al., *Classification and Nomenclature of Viruses. Fifth report of the International Committee on Taxonomy of Viruses,* 1991. Reprinted by permission of Springer-Verlag, Wein, Austria.

Fig. 18.3 Art: From R. I. B. Francki, et al., *Classification and Nomenclature of Viruses. Fifth report of the International Committee on Taxonomy of Viruses,* 1991. Reprinted by permission of Springer-Verlag, Wein, Austria.

Fig. 19.7 Art: Source: Data from C. P. Woese, *Microbiological Reviews,* 51(2):221–271, 1987.

Fig. 19.9 Art: From G. J. Olsen, C. R. Woese, and R. Overbeck, "The Winds of (Evolutionary) Change: Breathing New Life into Microbiology" in *Journal of Bacteriology,* 176(1):1–6, 1994. American Society for Microbiology, Washington, D.C. Reprinted by permission.

Fig. 19.9 Art: see public record.

Fig. 19.13 Art: Source: Data from G. J. Olsen and C. R. Woese, "Ribosomal RNA: A Key to Phylogeny" in *The FASEB Journal,* 7:113–123, 1993.

Fig. 29.11 Art: From John W. Hole, Jr., *Human Anatomy and Physiology,* 6th edition. Copyright © 1993 Wm. C. Brown Communications, Inc. Reprinted by permission of Times Mirror Higher Education Group, Inc., Dubuque, Iowa. All Rights Reserved.

Photographs

Fig. 30.6 Photo: © R. Feldman-McCoy/Rainbow.